BRITAIN IN THE NINETIES

BOOKS OF RELATED INTEREST

BRITAIN IN THE NINETIES

THE POLITICS OF PARADOX

Edited by

Hugh Berrington

FRANK CASS
LONDON • PORTLAND, OR

First Published in 1998 in Great Britain by
FRANK CASS PUBLISHERS
Newbury House, 900 Eastern Avenue
London, IG2 7HH

and in the United States of America by
FRANK CASS PUBLISHERS
c/o ISBS, 5804 N.E. Hassalo Street
Portland, Oregon, 97213-3644

Website: http://www.frankcass.com

British Library Cataloguing in Publication Data:

Britain in the nineties : the politics of paradox
  1. Political parties – Great Britain. 2. Great Britain –
Politics and government – 1979–
I. Berrington, Hugh B. (Hugh Bayard), 1928– II. West European
politics journal
320.9'41'09049

ISBN 0-7146-4880-9 (cloth)
ISBN 0-7146-4434-X (paper)

Library of Congress Cataloging-in-Publication Data:

Britain in the nineties : the politics of paradox / edited by Hugh
Berrington.
    p.   cm.
    Includes bibliographical references and index.
    ISBN 0-7146-4880-9. – ISBN 0-7146-4434-X (pbk.)
    1. Great Britain–Politics and government–1979–1997. 2. Great
Britain–Politics and government–1997- I. Berrington, Hugh.
DA592.B7415 1998
320.941'09'049–dc21   98-16038
                                                                CIP

This group of studies first appeared in a Special Issue on
'Britain in the Nineties: The Politics of Paradox' of
West European Politics, (ISSN 0140-2382) 21/1 (Jan. 1998)
published by Frank Cass.

Printed in Great Britain by Antony Rowe Ltd., Chippenham, Wiltshire

122198-4750I

Contents

Britain in the Nineties
The Politics of Paradox

HUGH BERRINGTON

The orator bustled up to him, and drawing him partly aside, inquired 'On which side he voted?' Rip stared in vacant stupidity. Another short but busy little fellow pulled him by the arm, and, rising on tiptoe, inquired in his ear, 'Whether he was Federal or Democrat?'
... 'Alas! gentlemen,' cried Rip, somewhat dismayed, 'I am a poor quiet man, a native of the place, and a loyal subject of the King, God bless him!'
Here a general shout burst from the bystanders 'A tory! A tory! ... hustle him! away with him!'

Washington Irving, *Rip van Winkle* (1820)[1]

The original Rip van Winkle went to sleep a few years before the American War of Independence, and awoke 20 years later to find when he returned to his village, that he recognised none of the inhabitants, that his wife was dead and his house in decay. The Union Hotel had replaced the old village inn; the sign that had borne the picture of King George III was still there, but though it still bore the face of the King, a sword had taken the place of the sceptre; the head had a cocked hat and the sign was labelled General Washington. Near to where the inn had stood was a flagpole, bearing the stars and stripes.

A modern Rip van Winkle, who went to sleep at the beginning of May, 1979 and awoke 18 years later, with just a short interlude of wakefulness in the summer of 1983, would have suffered the same kind of shock and endured the same bewilderment. He would have recalled the manifesto declarations on Europe in the two elections of 1974: Labour promising that it would 'immediately seek a fundamental renegotiation of the terms of entry' and the Conservatives asserting that 'Europe gives us the opportunity to reverse our political and economic decline. It may be our last.'[2] On waking, he would have seen the growing rift between the Conservative Party and big business, a division unprecedented in this century, on the very issue of Europe.

Before drifting off to sleep, he would have been aware of the rising turmoil within Labour, as its activists strove to seize hold of the levers of power; and he would have observed of the Conservative Party, that its one major, formal constitutional change since 1918 – the election of the leader by the MPs – had not yet revealed its full potential.

Had he awoken briefly in June 1983, before going back to sleep, his surprise on his final awakening in May 1997 would have been even more acute. He would have seen Labour go to the country in 1983 calling for Britain's immediate withdrawal from Europe, and its share of the poll reduced to 28 per cent, its lowest figure since 1918. He would have learned that the election of the Labour Party leader had been taken out of the hands of MPs and lodged with an electoral college.

As the rest of us adapt insensibly to change, it is only a Rip van Winkle who can savour fully the piquant smell of paradox. The contributions in this volume aim to explore in greater depth some of the paradoxes and contradictions of the last 20 years and their legacy for the present decade.

THE DISTRIBUTION OF POWER IN THE BRITISH PARTIES

Labour's turmoil, which began in the late 1970s, seemed like an epitaph on the work of the late Robert McKenzie. As Dennis Kavanagh shows in his absorbing review of the subject, McKenzie's seminal text, *British Political Parties*[3] first published in 1955, soon became the authoritative text for analysts of power within the two major parties. The distribution of power in the Labour Party had gradually come to resemble that in the Conservative Party, with Labour's democratic claims being slowly but remorsely challenged by what, for McKenzie, was an ineluctable process. The rift between the Labour government and the trade unions in the first Wilson government, however, posed serious doubts about the viability of the McKenzie thesis; the breakdown of the relationship between unions and government in the second Wilson and Callaghan governments, and the mobilisation of the constituency parties seemed fatally to undermine the thesis. Indeed, to many observers, the demands for constitutional change within the party seemed to pose a threat to the liberal democratic polity itself.

The changes proposed for the election of the party leader transferred the choice of prime minister from MPs to an electoral college in which they would have only 30 per cent of the votes. A larger share would be held by the trade unions, with the casting of those votes often influenced by horse-trading, and the balance by the constituency parties, often regarded as self-appointed, unrepresentative, remote from day-to-day experience of government and representative of opinion at the extreme of the political spectrum; the power to choose the prime minister would be placed firmly outside the House of Commons. Moreover, in government, the mechanisms to bring about the dismissal of the prime minister could be invoked by a two-thirds majority at the party conference. The changes in the Labour Party seemed to push the the constitution to the extra-parliamentary zone; the chain that linked the citizen, the House of Commons and the government would be broken.

Even at the time, such fears were crudely alarmist; in Canada, for example, party leaders are chosen by party conventions, whose members for the most part come from outside the House of Commons.[4] Similarly, in Germany the party leaders, appointed to head their parties during the election campaign are chosen by the party organisations outside the Bundestag. None of the consequences foreshadowed for Britain have occurred. Indeed, the remarkable feature of the last 15 years has been the ease with which the new formula for electing the Labour leader has been absorbed into normal constitutional practice.

The second procedural change, the mandatory re-selection of MPs, has also not had the effects that its proponents had hoped for and that its enemies had feared. There were sundry casualties and would have been more numerous but for the timely defection of some MPs to the newly-formed Social Democrat Party (SDP). But there was no mass purge and today most Labour MPs who wish to stand again are re-selected as a matter of course. Again, the constitution has withstood the tremors aroused by this innovation.

In retrospect, what stands out about Labour's constitutional changes in the early 1980s is the facility with which they have been reversed or neutered. At the time, the struggle did not seem easy, but with hindsight the losers of 1979–81 have become, vicariously, the victors of the 1990s. Looking at Labour's history since 1945, not only at the constitutional struggles, but also at the debates over policy, the most striking feature is the way in which the aims of the revisionists of the 1950s have been achieved, not by the old Gaitskellite right of the party, nor even by Wilsonite trimmers, but by the soft left. Gaitskell failed in his bid to amend Clause IV; but this was carried through over 30 years later by the one-time member of the Tribune Group, Tony Blair.

The Conservative Party is following Labour, but at some distance in time. William Hague has already conceded the direct engagement of members in the selection of the party leader. It is only 34 years since the Conservatives chose the last leader to be nominated by the so-called Magic Circle. The adoption of a formal election by MPs in 1965 marked a sea-change, which gave and arguably restored to MPs (albeit to a section of them) the function of choosing and dismissing governments. However, the choice was lodged in the hands of MPs alone, with the party organisation playing only an advisory (and largely disregarded) role, its view elicited in a manner worthy of the Magic Circle. Within a few months of the election defeat, the right of party members outside the House to have a vote in the election of the leader has been accepted; all that remains is to determine how much weight they will have.

Walter Bagehot, writing in the wake of the Reform Act of 1867, which he opposed, observed that the outcome of this prime constitutional change would probably depend on the character of the issues that the country's statesmen presented to the electorate. If Bagehot grieved over the extension of the franchise, he also thought that something might be retrieved from the nation's plight. It was the task of political leaders to formulate the issues which the new electorate would resolve; on the care they took would hang the results of a delicate experiment.[5] Despite occasional lapses, control, since 1867, of the electoral programmes of the parties and the manifestos submitted to the electorate, has rested with the parliamentary leaders of the great political parties. Even at the time of the constitutional revolution of the early 1980s, the Labour leadership kept effective sway over the party's manifesto. The activists' charge that responsibility for the election manifesto should be lodged with the National Executive Committee (NEC) was the only demand, of the three presented, not to be supported by the party conference. Labour's submission of its election programmme to the party membership, on a take it or leave it basis, was a figleaf to hide the dominance of the party leadership. The Conservatives, under their new leader, have promised to consult party members about the contents of the election manifesto. Again, sceptics may reasonably doubt whether the grip of the parliamentary leadership will be dislodged. But both Labour and the Conservatives have enfranchised, or are about to enfranchise, their members outside Parliament in the choice of leader. We may doubt whether these members will be able to impose their candidates on unwilling MPs. Formally, however, both parties have opened the gates to the extra-parliamentary parties.

These changes cumulatively represent an acute challenge, at least in form, to traditionalist interpretations of the constitution. They afford a major re-definition of the formal sources of executive authority: whether they stretch beyond a new definition and will alter the substance of policy is much more doubtful.

THE PARADOX OF EUROPE

The present stance of the two parties on Europe affords a striking paradox.

'For the greater part of my political life', said David Curry when resigning from the Shadow Cabinet, 'Conservatives have been the party of constructive engagement in Europe and Labour the party of isolation.'[6] It would be hard, over the last hundred years, to find any parallel in British political history.

Two similarities emerge from the accounts of the rise of Euroscepticism in the Conservative Party (Berrington and Hague), and Labour's conversion to the cause of European Union (Daniels). In both parties, the onward march of the generations in the parliamentary parties helps to explain the dramatic turnarounds in policy. The parliamentary party is a more important theatre for the enunciation of policy in the Conservative Party than for Labour, and the change in composition has had a more explicit effect among the Conservatives than it has in the Parliamentary Labour Party.

Generational changes, though, are easier to record than to explain. The long Conservative ascendancy and Thatcherite economic and social policy, coupled with greater sympathy for social measures in the European Union itself, help to elucidate the growing support for Europe among the younger Labour backbenchers. The grass began to look greener on the other side of the Channel.

The replacement of Edward Heath by Margaret Thatcher may well account for some of hostility to closer European integration shown by the more recent intakes of Conservative MPs, just as it might explain a similar reversal among ordinary party members. It seems doubtful, though, that a single change of this kind can fully account for the heavy switch in sentiment to be found amongst the more recent cohorts of Conservative MPs. Norris and Lovenduski's survey of MPs and candidates, conducted during Margaret Thatcher's last, and John Major's, early months in office did not foreshadow any major swing in attitudes to Europe.[7] The same evidence seems to play down the direct significance of changes in the selection of Conservative candidates. The likeliest explanation is that there had been a latent evolution of opinion amongst the more recent cohorts, which crystallised into scepticism or whole-hearted opposition after the signing of the Maastricht Treaty. Suddenly, many who had seen closer union with Europe as part of the conventional wisdom began to have doubts.

Among both parties, the question of sovereignty has loomed large in antagonism towards closer links with Europe. In the Labour Party concern for sovereignty, at least in public discourse, has been almost wholly instrumental; the preservation of sovereignty has for most never been an end in itself. Sovereignty made possible the implementation of distinctive socialist policies. British socialists could only enter the promised land of socialism through the gateway of the nation state. Thatcherites shared the same millenarian zeal for their cause, but they were joined in opposition to Maastricht by numerous others for whom a sovereign Britain is an end in itself.

On the Labour side, Phil Daniels' illuminating account shows that there was no clear point of origin in Labour's *rapprochement* with the European Union. Paradoxically, the new attitude to Europe was a by-product of the change in its economic policy which began after the electoral defeat of 1983: 'a commitment to Europe was a key part of Labour's efforts to enhance its credibility as a party of competent economic management'. But it was the conference vote in 1989 in favour of a cautious welcome to Europe which finally legitimised the new shift in policy. By that year, the very constraints on a Labour government's economic strategy, which before 1983 had aroused such indignation on the party's left, were now seen as providing a welcome screen for any Labour Cabinet exposed to pressures for an expansionary programme.

Whatever the cause of the reversal in feeling towards Europe in the Conservative Party, the issue affords a conspicuous and unusual example of that party changing policy as the result of pressure from below. The hierarchical traditions of the party, the special place assigned to the Leader by party mythology, stand in contrast to the way the Eurosceptics captured the party. What began as a backbench rebellion ended with the election of a new leader pledged to reject the single European currency, and the party's repudiation of its former chieftains.

THE CHANGING FACE OF THE CONSERVATIVES

The European issue shows that the change in the character of the Conservative Party has been more important and more striking than the changes in the formal rules which govern the distribution of power in the parties. The behaviour of the party, certainly during the last ten years and arguably for the last 20, has defied all the comfortable stereotypes of the McKenzie era. The common social backgrounds, the similar school, university and regimental experiences and the informal connections between families provided both common standards of behaviour and a network of communication. So, the code of conduct in Parliament, or among the leaders in the constituencies, meant that disagreement was rarely publicly voiced, or if voiced, not stridently so; the informal channels afforded a route between party members and the party leadership; amendments to policy could be put discreetly and concessions offered quietly. Controversy within the party in the era of consensus was rarely prompted by ideology and often had the character of quarrels about administrative detail. The public face of the party was one, not merely of outward unity, but of inner harmony. The expression of political dissidence was governed by the code of 'the done thing'.

This was certainly the image by the Conservatives projected in the 20 years after the end of the war. What, in retrospect, is so striking about the Conservative disarray at the time of Suez is not that it happened, but that it was resolved so quickly and with such little fuss. The same applies to the battles between backbenchers and government towards the end of the Macmillan era. The restraints that the Conservatives observed both in the House and in the constituencies made almost redundant any formal scheme of discipline; until recently, it was more common for members to renounce the whip themselves, as a gesture of disapproval of the party leadership, than for the leader to withdraw the whip as a sanction. John Major's decision in 1994 to withdraw the whip from eight Eurosceptic MPs who voted against the European Communities Finance Bill, was without precedent in its scale and suddenness.

What has changed in the last two decades has been the style and manner of dissidence in the party. Discourse in the party no longer has that quality of civility. In the past, the party sought behind closed doors to clarify its disagreements and resolve its differences: 'loyalty' said Lord Kilmuir, 'was the Tories' secret weapon'.[8] When it needed to, the party leadership made adjustments in policy with little publicity, and with the minimum of recrimination.

Indeed, the stiff upper lip was institutionalised in the proceedings called formally to ratify the choice of Conservative leader, after the decision had been made through the so-called customary processes. The defeated rival, or perhaps a long-standing critic of the new leader was called upon to propose the election of the choice which had 'emerged'; so Lord Curzon had to hide his disappointment in 1923 and endorse the name of Stanley Baldwin, while Churchill, as seconder, had to perform the same office in 1937 for Neville Chamberlain.[9] This practice found a parallel in the nomination of the President of the USSR when Mikhail Gorbachev proposed the election of Mr Chernenko,[10] who had already been chosen as General Secretary of the Communist Party.

The change in style developed gradually. It clearly owes something to the introduction of a formal procedure for electing the Leader. It became more manifest in the closing years of Mrs Thatcher's premiership, especially at the top; the public rancour displayed in the disputes between Mrs Thatcher and two of her leading ministers, Nigel Lawson and Sir Geoffrey Howe, was unprecedented since the war. In one instance these frictions were aggravated by mutual dislike – 'we found each other's company intolerable', as Mrs Thatcher wrote of Geoffrey Howe.[11] But the resignations of two of her former chancellors were but a foretaste of what was to come under John Major. The cabinet infighting, Mr Major's own comments about some of his ministers (cabinet bastards), the openness with which Europhile and Eurosceptics assailed one another and the public demonstrations of dissent in the House of Commons were all new to a country whose Conservative Party had for so long been a watchword for stability and good manners. Even the early publication of memoirs, like snakes' eggs venomous at least in parts, testify to the emergence of a different kind of Conservative Party. Nor is there any sign of a reversion to the civility of an earlier period. Mr Hague's apology for Britain's adhesion to the European exchange-rate mechanism at a conference attended by the man who had made the decision and who was his predecessor as leader, stands alone. Even the poll tax was buried, at dead of night. 'Few and short were the prayers we said, And we spoke not a word of sorrow.'[12]

Our Rip van Winkle would have wondered whether Conservative fratricide was something borrowed from Labour; he would have witnessed the hatreds unleashed in that party at the end of the Callaghan government and might have surmised that the infection had spread to the Conservatives. Anthony King's seminal article on the rise of the career politician in Britain, predicted consequences not unlike those we find in the Conservative Party today.[13] A parliament composed of Career Politicians, of ambitious men and women keen to make their mark, lacking the ballast of the passive member or the part-time representative, will exacerbate competitiveness and nourish ill-feeling:

> ... a legislature containing a high proportion of career politicians is likely to be a restless, assertive institution. ... More backbench MPs want office, but there are not nearly enough offices to go round. So, frustrated they seek other outlets for their energies and self-assertiveness.[14]

King goes on to attribute a decline in the collegial character of decision-making at the top of British politics to the influx of ambitious, full-time, professional politicians. Perhaps the difficulties faced by the contemporary Conservatives reflect the departures of the knights of the shires, the eclipse of the gentlemen and the rise of the players.

But Rip van Winkle would then gaze at the Labour Party. Could this be the party he knew, whose malice defied containment and whose resentments were flaunted with such élan. The same tendencies in parliamentary recruitment were taking place in both parties; the old working-class trade unionists, like the Tory knights of the shires, were gradually giving way to an influx of young, university-educated professional politicians. Why has Labour been able for so long to hide its conflicts and suspend its vendettas?

The answer may simply be that history has not yet written its verdict on New Labour, that the thirst for office and the euphoria of its landslide victory have so far submerged the divisions between right and left. Let the government face economic adversity, or a sharp fall in popularity, and perhaps the latent ill will will be revealed.

A political party, perhaps mercifully, lacks a collective memory. The behaviour of Conservative ministers and MPs might puzzle Rip van Winkle, but not if he had recall stretching back across the generations. For however decorous the Conservative Party has been in the handling of its controversies since the end of the Lloyd George era, the early years of the century showed all the antipathies which came to life in the Major era. Wholly in spirit, and often in detail, the factions in the Labour Party in the 1970s and 1980s harked back to the example of the Unionists in the Edwardian age. Deselection, for instance, was not a practice invented by the Labour left. It was the Unionists in the controversies over Tariff Reform who set the precedent for deselection of members on policy grounds, who published blacklists of colleagues to be opposed at elections, who set out to capture constituency Conservative associations.[15] Baker, Ludlam and Gamble point to the example of tariff reform, and before that corn-law repeal; in both, as over Europe, the question that divided the party was the issue of Britain's economic relations with the outside world. Is there something about the nature of these issues that accounts for the frequency of rebellion and the severity of the problems facing Conservative Party managers? Perhaps so, but can it also explain the virulence and the acrimony which have so tarnished relationships among colleagues?[16]

THATCHERISM

In their elegant essay on Thatcherism, Bevir and Rhodes try to tease out the meaning of the term, in the light of four traditions in British politics – Toryism, Liberalism, Whiggism and Socialism – and bring out some of the paradoxes of Thatcherism. They conclude that there is no single tradition with which Mrs Thatcher can be fully identified and indeed that there is no essentialist account of Thatcherism. Perhaps however, Thatcherism is best understood by reference to a tradition that owes far more to France than to Britain. The tradition that comes closest to Thatcherism is Jacobinism, with several significant parallels. English individualist radicalism of the mid-nineteenth century is, historically, the nearest creed we have to an indigeneous Jacobinism. Its leading features, as delineated by John Vincent in his brilliant essay in *Pollbooks: How Victorians Voted*, were its opposition to public expenditure and its hostility to authority. In the small towns and villages of mid-nineteenth century England, authority took the form of local notables and, more broadly, the aristocracy. In the emerging collectivist society, we may add, the bureaucrat and the inspector were to become the embodiment of authority. Vincent describes the phenomenon as English Jacobinism, and refers to the puzzling 'resemblance between French and English radicalism in the later nineteenth century'.

The social base was provided by the self-employed – the artisans and the shopkeepers; that social base 'precluded any economic programme except economy'.[17] The cobblers of Northampton were one of the most vehemently radical groups in the country, and chose the ultra-radical Labouchère and the atheist republican Bradlaugh in election after election. As late as 1888, Bradlaugh and his colleague Labouchère could attack a bill to limit the hours of labour of shop assistants.[18] Individualist radicalism died out as a single, recognisable doctrine but entered the bloodstream of both the Conservative and Labour parties, each absorbing different elements. Hostility to high taxation and economic regulation found a welcome home among the Conservatives, while the distrust of armaments and militarism, and the programme of democratic reform found a haven in the newly-born Labour Party. So, Tony Benn's uncle, the publisher Ernest Benn, author of *The State the Enemy*, founded the Society of Individualists, while his father, William Wedgwood Benn, left the Liberals to join Labour.

Much has been written of the influence of Margaret Thatcher's father, Alfred Roberts, on her political creed. The world has taken little note of her paternal grandfather, a Northamptonshire cobbler. Margaret Thatcher's ideology has many points of resemblance to the economic beliefs of the cobblers.

English individualist radicalism contained sundry features which in the modern context might be regarded as akin to Jacobinism; paradoxically, however, Thatcherism, as it developed, has more affinities with the French than the English version. The resemblance would doubtless be repugnant to Mrs Thatcher who wrote:

> The French Revolution was a Utopian attempt to overthrow a traditional order ... in the name of abstract ideas, formulated by vain intellectuals, which lapsed, not by chance but through weakness and wickedness, into purges, mass murder and war. ... The English tradition of liberty, however, grew over the centuries: its most marked features are continuity, respect for law and a sense of balance ...[19]

However, distasteful though the comparison may be to Mrs Thatcher herself and her votaries, the points of likeness are numerous. Her mission, to roll back the frontiers of the state, has historically been the creed of the self-employed and small business. In her landslide victory of 1983, the petty bourgeiosie proved to be her staunchest backers. Seventy-one per cent voted Conservative, a far higher proportion than of any of the other four strata identified by Heath, Jowell and Curtice. Yet, it was not simply that this class (the earnings of whose male members ranked lower than those of the male working-class sample) voted overwhelmingly for the Conservatives; their attitudes, on issues such as income redistribution, nationalisation and job creation, were much farther to the right than those of any other group.[20]

Mrs Thatcher's populist contempt for the 'establishment', and her abrasive disdain for inherited privilege, made her an unusual Conservative leader. She made clear her preference for the self-made, the meritocrats and the outsiders – and Jews, so long an outsider group, were prominent in her cabinets, an attribute summed up in the tasteless remark attributed to Harold Macmillan: 'Nowadays, there are more Estonians in the Cabinet than Etonians.' The bureaucracy was another target that fitted closely with the stereotypes of English individualist radicalism 'there was no doubting Mrs Thatcher's hostility to much of the ethos of the civil service.'[21] Like the French Jacobins, she attacked the partial associations: the trade unions, the civil service, the universities and later, professions such as the law; and upbraided others such as the churches. She assaulted the intermediate institutions: in the Britain of the 1980s, this meant local government. The metropolitan counties were summarily abolished, the remaining local councils were subjected to stringent financial controls and their most important functions were either reduced in scope or taken entirely away from them. Sometimes these functions were given over to appointed quangos, sometimes to *ad hoc*, otherwise unorganised groups of consumers or parents. The latter transfer of functions may or may not have been a form of giving power to the people, but it did represent a diminution of the importance of the intermediate bodies. Her onslaught on local representative institutions was characteristic of the French Jacobins. The kingdom was one and indivisible; its sovereignty could not be sundered either by powerful local councils or assemblies with devolved powers.

These themes came to a grand unity in her defence of the British nation-state. She saw the nation-state as the ultimate form of political organisation. She rejected union above, as well as devolution below.

Nor does the similarity end here. Jacobins proclaimed their message with messianic enthusiasm and sought to export it to the rest of Europe. Indeed, one prime incentive to resist the European Union's thrust towards federalism lay in the need to safeguard the purity of the Thatcher experiment. If it failed at home, it could not prevail abroad.

Margaret Thatcher's revolution, therefore, was a Jacobin revolution, even though the handbag supplanted the guillotine. It was a revolution carried out through the agency of a Conservative majority and in a social context far removed from that of pre-industrial France. Every Conservative leader has to reconcile the divergent, sometimes conflicting, traditions which the modern Conservative Party has inherited, and to do so in a way which meets the exigencies of the modern world. Mrs Thatcher herself notes a 'curious discrepancy' in the behaviour of her ministerial colleagues of the 1960s and 1970s:

> What they said and what they did seemed to exist in two separate compartments ... But the language of free enterprise, anti-socialism and the national interest sprang readily to their lips, while they conducted government business on very different assumptions about the role of the state at home and of the nation-state abroad.[22]

The contradictions within Thatcherism will long continue to fascinate observers of British politics. Her unswerving nationalism and her economic individualism lie uneasily together; her resistance to the European Union's incursions of sovereignty contrast with her readiness to follow the lead of the United States. Moreover, she was loath to acknowledge that markets and regulation are not always enemies and could not accept that the Single European Market, of which she was a partisan, actually demanded more, and indeed constant, regulation. Britain has been the biggest single source of European regulation, all in the name of creating and consolidating the market.

Bevir and Rhodes cite Willetts as averring that Mrs Thatcher's achievement was to reconcile Toryism and individualism. Issues of personal morality have posed particular dilemmmas for Conservatives, for here the individualist tradition drawn from liberalism (or, arguably, more specifically from radicalism) runs counter to the Tory emphasis on tradition and authority. So, the allegedly statist Labour party provided the Commons' votes which decriminalised homosexual behaviour beween consenting adults, while a majority of Conservatives went into the division-lobbies to vindicate the status quo.[23] Mrs Thatcher herself was one of the earliest Conservatives to vote for the legalisation of homosexual behaviour,[24] an action demanding considerable courage at the time, though her vote is rarely recalled by her admirers; her zeal, however, for this kind of moral libertarianism does not seem to have lasted.

It was an easier task to reconcile authority and self-interest; their accommodation was one of the achievements of Thatcherism. The hidden hand and the strong state combined to bring the two together. Yet, the Thatcherite programme was not a consciously articulated catalogue of measures, designed to set the people free; it resembled more the nineteenth-century letter, sent by the party leader to his constituents than a modern manifesto. Rather, as Riddell says, as quoted by Bevir and Rhodes, Thatcherism was 'essentially an instinct' not an ideology, or again to use Riddell's words, 'a rolling policy agenda', as the example of privatisation testifies. Furthermore, say Bevir and Rhodes, there is no single notion to be explained. 'Thatcherism ... was ... rather several overlapping but different entities constructed within overlapping but different traditions.'

THE REGULATORY STATE

Mark Thatcher relates, with lucid economy, the dilemmas posed by the nationalised public utilities: 'the decision-making processes 'were informal and involved discussions and negotiations conducted in private ... policies were rarely consistent as one set of short-term pressures and constraints succeeded another'. The boards were supposed to represent the public interest, with the minister as the ultimate guardian, but 'the interests of the utility suppliers and manufacturers were at the forefront of policy'. Collectivist opinion deemed the industries, as natural monopolies, as being ripe for public ownership and supervision, but evaded the quandaries entailed in giving flesh and blood to the term 'the public interest'.

Privatisation might symbolise the triumph of neo-liberal doctrine over Fabian socialism; yet, it was essentially an unplanned, almost incidental, victory and not surprisingly the terms on which the new policy was introduced lacked detailed clarity. 'The position of the industry DGs [Directors-General] under the various statutes is ambiguous... The development of the position of the DGs was not foreseen at the time of their creation.' The riddle that the nationalised industries never solved – how to identify and promote the public interest – was passed to the DGs of the newly privatised utilities. As with most British institutions, what success they have had has come, not through the invocation of an abstract formula, but by pragmatic responses to the exigencies posed by experience.

Their work is best known to the public through the imposition of price controls; the apparent greediness of the boards, and presumably of the shareholders, of the new companies, provoked widespread public anger. Those who can recall the criticisms of the nationalised industries in the late 1940s and early 1950s – high prices, poor service and insensitivity to consumers – will have experienced a delicious sense of paradox at the public response to privatisation, referred to by Denver.

If Mrs Thatcher and her colleagues pushed back the boundaries of government, through privatisation and disengagement, the liberated provinces did not enjoy their freedom for long. What is striking about the action of the regulators on prices is the confidence with which they have imposed their decisions and the autonomy they have enjoyed so far. But although it is the actions of the DGs to restrain prices that have led to the newspaper headlines, their work has been far more extensive. Their pursuit of 'fair and effective competition' has had profound implications for the industries, as well as providing an example of how regulation can be pursued.

The achievement of the regulators prompts memories of the arguments on the left in the 1930s about whether it was necessary for the state to own an industry in order to control it. The model offered by the regulators could be a pattern for a future left-wing government. Indeed, as Thatcher observes, the 1980s and the 1990s have not seen de-regulation or the famous rolling back of the frontiers of the state. The new regulatory regime has actually strengthened the state; it 'has led to a more powerful state than the previous arrangements under nationalisation'. But the result is a state with a difference; if, to the privatised utilities, the DGs 'often represent the ever present and over-powerful state', it is a more fragmented state. The capacity of governments may have declined as a result of these innovations; the capacity of the state may have increased. To the privatised toad beneath the harrow, the state looks all-powerful; the butterfly flying above the road is more impressed by the fragmentation.

THE ELECTIVE DICTATORSHIP?

The concentration of power in Britain has provoked the description of the political system as an elective dictatorship. The term is usually attributed to Lord Hailsham, in his book *The Dilemma of Democracy*, published in 1978.[25] The notion, however, was developed by R. W. K. Hinton, as long ago as 1959, in his article, 'The Prime Minister as an Elected Monarch'[26] (the term monarch being used in the literal sense of single ruler). Hailsham's phrase emphasises the power of the executive above the other branches of government. What was new was not so much what he said, which was commonplace, but who said it, and the elegance with which the thesis was propounded. An election confers a majority, and by virtue of the First-past-the-post formula, usually an exaggerated majority, on one or other party, whose leader becomes prime minister and appoints the cabinet. The cabinet, through the working of party discipline and its control of the timetable of the House of Commons, is able to carry its proposals to the statute book largely unscathed. To the statute book and beyond; for the lack of any limitation imposed by a written constitution means that the courts must accept any Act which has passed through parliament in the prescribed way and received the royal assent. The measures a cabinet proposes cannot be tested against the yardstick of a bill of rights, still less mitigated by an entrenched written constitution; they cannot be invalidated by appeal to a superior law. The majority party holds untrammelled sway for up to five years, when it meets the electors again. If defeated, it is replaced by a new elective dictatorship.

This sketch was, at best, a parody of the working of the constitution in the era of consensus. It ignores, for example, the positive virtues of pressure groups; Lord Hailsham acknowledges the existence of the trade unions and of single-cause groups, but sees the former not as a possible curb on the executive, but as a branch of the dictatorship which is part of the problem, while the cause groups are seen to derogate from rational and civilised debate within Parliament. It assumes that the majority party is united and homogeneous and that its leaders never anticipate the reactions of the electors or of a crucial section of them.

Proposals to temper the supposed power of the elective dictatorship relate either to a strengthening of the legislature against the executive or to the adoption of a written and arguably entrenched constitution; the latter would not only weaken the executive, but also qualify the formal powers of the legislature.

The former alternative, the strengthening of the legislature, has already occurred under the Conservative governments, in the shape of the comprehensive structure of select committees since 1979, though the limitations on the powers of these committees are considerable. Moreover, as Philip Norton has shown, the near 100 per cent conformity once displayed by backbenchers in the division lobbies has been diluted. While it is easy to exaggerate the significance of the greater freedom claimed by backbenchers, and indeed to overestimate the independence of backbenchers of a hundred years ago, there has been a perceptible increase in members asserting their independence of the whips[27] (a development which has not diminished public criticism of the role of party as a straitjacket upon MPs).

However, the biggest limitation on the power of the 'elective dictatorship' has already been mentioned. The formal election of the party leader in the Conservative party has introduced a new element into the relations between leaders and followers. The provision for the annual re-election of leader may have been intended as a formal ritual, not to be invoked when the party was in office. The example of Mrs Thatcher, and the travails of John Major have shown that this clause is no dead letter. Any Conservative leader must have his or her eye on the possibility of a challenge each November. The incorporation of the representatives of the grassroots may, ironically, as some critics of the proposal have indicated, fortify the position of an incumbent leader.

One of the paradoxes of almost 18 years of Conservative government, at first sight, has been the apparent strengthening of central government at the expense of local authorities and social institutions. One strand of right-wing thought has emphasised the dangers of over-powerful government and the importance of independent local government and a social diffusion of power, as ways of correcting the concentration of power within the executive, and geographically, at the centre.

THE JUDICIAL DIMENSION

The Conservatives' long tenure of power seemed to substantiate the charges of elective dictatorship; Conservative backbenchers' occasional expressions of dissidence in the division-lobbies only nibbled at the edges of executive power, as did assertions of independence by the House of Lords. More notable were the interventions of the courts which have led to what Nevil Johnson calls a double paradox: the party of individualism found itself baulked by the courts, and the party of collectivism welcomed the judicial intrusion. Yet, Johnson's subtle appraisal finds the cause, procedural simplification apart, to lie as much in the development of a rights-based culture (with increasing litigation by pressure-groups) as in the desire of judges to affirm the need for limits on the power of the executive.

The issues raised in Johnson's chapter on the courts are acute and go to the heart of the perennial debate between the advocates of popular rule and the champions of limited government. The argument may be perennial, but the placement of the disputants presents a paradox, visible in the grudging reaction of Conservatives to the present government's proposals for the incorporation of the European Convention on Human Rights into British law. While Conservatives propound the doctrine of parliamentary sovereignty, and hence indirectly of majority decision, Labour (while resorting to some procedural ingenuity to protect legislative sovereignty) espouses the need for judicial checks. Nevertheless, however paradoxical the positions of the parties may be on judicial activism, they are not without precedent. Nor is this altogether surprising; there has always been more to the Labour Party than Fabian collectivism. If part of the nineteenth-century tradition of individualist radicalism was absorbed by the Conservatives, re-surfacing in the Thatcher era as neo-liberalism, part entered Labour's mainstream and was reflected in the defence of the subject against abuses of power by the traditional authorities. It was the Labour government of 1945, which to many seemed to represent the apogee of collectivist intervention in economic and social life, which introduced the Crown Proceedings Act.

More pertinently, it was Harold Wilson's government which brought in the office of the ombudsman in 1967. Significantly, support for this proposal increased towards the end of a long period of one-party rule. Although the bill went through parliament without difficulty, the reaction of Conservative ministers beforehand had been both dismissive and defensive. It is still possible to recall ministers explaining in television interviews how unnecessary such an office would be. The creation of a new office would cut across the normal lines of ministerial responsibility. The minister himself was the citizens' true ombudsman; alternatively, the backbench MP was the appropriate guardian of the citizen against the bureacracy.

The decision to incorporate the European Convention of Human Rights into British domestic law, is likely, as Johnson observes, to strengthen the role of judges and to provide a further instalment of the rule-bound mode of government. The same applies to the proposed Freedom of Information Act. Again, it is significant that support for both measures has accumulated at the end of a long period of one-party rule. The bill to incorporate the Convention into British law will, however, stop well short of empowering judges to strike down legislation which they see as negating the rights protected by the Convention.

Johnson has three main fears about the present 'rush to legalism'. Legalism diminishes the discretion available to ministers, for the exercise of which they are accountable to the House of Commons. The leaven of flexibility in policy will be lost. Moreover, an increase in the powers of judges means a corresponding reduction in the powers of politicians and, ultimately, of the electorate. Finally, he warns of the paradoxical consequences of 'binding judicial interpretation ... in the name of equal rights and freedoms for all'. It may be said, too, that at the base of the demand for judicially enforceable human rights is the view, often only semi-articulate, that human rights form part of what is in essence a positive sum game. There are the people and – what from this standpoint can be properly ignored – the political élite. The people's rights can be extended by simply encroaching on to the territory jealously guarded by the élite.

Such a simple opinion is the 1990s version, cast in judicial form, of the view widely proclaimed in the 1960s that greater political participation by citizens, through the exertion of pressure on governments, could extend human freedom almost indefinitely. There seemed no realisation that the more the various groups mobilised to exercise political pressure, whether through the vote or other means, the less discretion governments would have and the less capacity to respond to citizens' grievances. The very term 'rights' suffers from being overworked; in what has been called 'the escalation of rights rhetoric'[28] the term has been stretched to include almost everything seen by individuals, or even society, as being highly desirable. An inflation of terms simply cheapens them and must diminish the force of genuine rights in ordinary social discourse. In mature, democratic societies there can be no simple dichotomy between people and élite. The people represent a congeries of minorities, which sometimes cut across one another, and which are sometimes isolated and distinctive. In the end, an evolution in the direction of judicially-enforceable human rights 'may benefit a disparate range of minorities at the expense of diminished freedom and responsibility for the majority' (or, it might be added, other minorities). After 25 years it is still possible to remember trade-union and Labour criticism of the Heath government's trade-union legislation that it was not appropriate to bring the law into industrial relations; and the controversy over the rights of homosexuals to serve in the armed forces, despite, it seems, the bitter opposition of many serving soldiers, calls to mind the the voice of trade unionists in the arguments about the closed shop: 'Don't we have the right not to work with these men?'

The problem is that the rights of some categories of persons – ethnic and religious minorities, asylum seekers, beggars, paedophiles, other criminals, the psychologically disturbed – raise some of the most sensitive and indeed potentially inflammable issues in British politics. It requires no great constitutional dexterity to safeguard the rights of popular minorities; an entrenched constitution is hardly needed to make them secure. The purpose of having special constitutional procedures is to defend the rights of unpopular sections of the population. In so far as such measures are successful, there is the risk of acute public feeling, unable to express itself to the relevant decision makers, spilling over into unconstitutional or violent action. The way Parliament takes some major decisions, without recourse to the people through a referendum, apppears to excite widespread public irritation. How much more will the decisions by judges, invulnerable to public rebuke do so?

Nevetheless, there is evidence to show that concern over civil liberties is more than a fad of academic jurists and constitutional pundits. Elsewhere, Miller, Timpson and Lessnoff report a general distrust of politicians among the sample they interviewed in 1991–92, and it is, therefore, not surprising that a bill of rights incurred considerably more public approval as a protector of citizens' rights than did backbench MPs. When respondents were asked whether the final say in a dispute about a law should be given to the British courts, or to Parliament, only a 'slender majority' said Parliament; but no fewer than 55 per cent would allow the European Court to have the last word. Such a finding is clearly counter-intuitive, but it corroborates other surveys reporting a sharp decline in confidence in traditional British institutions.[29]

THE PERIPHERY

The politics of the periphery is rich in paradox. William Miller shows
ironies jousting with absurdities and quarter-truths tilting with
contradictions. There is the spectacle of the Labour Party, which following
its Fabian head repressed its Home Rule instincts in 1958, now
spearheading the thrust for devolution; and perhaps more richly of the
Conservative Party led by Mrs Thatcher declaring that a devolved Scottish
assembly must be a top priority.

There is nothing new under the sun, and the West Lothian Question, posed
by Tam Dalyell, does no more than raise again the issue broached by the status
of the Irish members at Westminster, in the Home Rule Bill of 1892 (and
indeed of 1886). Should they be excluded for all but 'one or two special
purposes', included for all purposes, or included for imperial purposes only.[30]
William Miller shows (like a philosophy paper in a finals exam), that the often
asked West Lothian Question, which he renames the Catalan Question, is not
a question – or, perhaps, more accurately has not yet been formulated as a
question. It is, instead, 'a frivolous debating slogan'.

Under Mrs Thatcher the Conservative and Unionist Party re-discovered
its Unionist origins, after Mr Heath's dalliance with devolution. For her
successors, John Major and William Hague, the one and indivisible
kingdom has remained and is to remain the exclusive focus of political
loyalty. It is hard to see why Conservative opposition to devolution is
voiced with so much passion: if devolution poses a risk to the Union, by
providing through the new Scottish assembly a sounding-board for
separatism or by fostering conflict beween Edinburgh and Westminster, so
does doing nothing. Any form of local or regional institution may afford a
focus for the sentiment of autonomy. But whether or not it is more
dangerous for supporters of unity to concede such institutions than to go on
resisting, is essentially a matter of judgment, hardly a basis for the
apocalyptic prophecies from the Conservative front bench. Moreover, it
seems perverse to deny that focus when there is such a clear feeling of
singularity. 'You can have your council, or assembly, or parliament,
provided there is no long-standing, natural and distinctive sense of
community to justify one.' The injunction of Edmund Burke, 'to consider
the wisdom of a timely reform', is as apt to the constitutional relationships
of the four countries as it is to any other contemporary issue.[31]

Thatcherism has raised, in the starkest way, the issue that is at the heart
of the current constitutional debates, that of sovereignty. Or – has it? Miller
cites Enoch Powell 'power devolved is power retained'. If so, what is all the
fuss about? Presumably, opponents of devolution do not accept a lawyer's

definition of sovereignty and see the capacity of the British Parliament to act as being in some way compromised by the grant of devolved powers to Scotland. To aver this view risks undercutting the whole of the argument based on legal definitions of sovereignty. The power of a supposedly sovereign parliament to act is qualified in all sorts of ways, regardless of its legal omnipotence. This indeed is the kernel of the argument offered by British supporters of closer European Union. Globalisation of the economy limits the leeway of British governments seeking to maximise Britain's economic strength. To such, the loss of sovereignty is not a real loss at all if the powers relinquished cannot in practice be independently exercised. As Harold Laski put it in 1938: '... there is, historically, no limit to the variety of ways in which the use of power may be organised. The sovereign State, historically, is merely one of those ways, an incident in its evolution the utility of which has now reached its apogee.'[32]

In the same way, the tax-raising power, the subject of so much contention, is shown to be another debating catchword. The tax-raising power is unlikely to be invoked, even after the five-year period of grace conceded during the election; and changes in the allocation of the block grant offer a Scottish assembly far more scope to alter policy than the 3 pence tax-raising capability.

Subsidiarity, the watchword of those seeking to limit the exercise of the powers conferred on the European Union by the Maastricht Treaty, is a two-edged sword. If Britain can reasonably claim that powers are better exercised by the British government and parliament, than by the European Union, the same can be said of a Scottish assembly or even local councils. From the standpoint of believers in the sanctity of the nation-state, as Miller observes, the concentration of all powers at the level of the nation-state, by taking powers from levels both above and below the state, is not at all absurd. 'The paradox is merely that some Tory Eurosceptics have used the term "subsidiarity" to describe what is its exact antithesis: in the name of subsidiarity they have tried to concentrate power in one institution at one level.' To this we may add that this concentration of power runs against the grain of decades of Conservative rhetoric extolling the local, the small-scale and the informal against the remote, the extensive and the bureaucratic.

THE ELECTORATE

The election result, as David Denver's revealing analysis shows, displayed two great paradoxes. Labour won, and won overwhelmingly, despite a considerable economic recovery. Labour also won despite a long-term decline in the size of its working-class base.

The long success of Mrs Thatcher's Conservative party fostered a mood of pessimism within Labour and among sympathetic commentators, which was compounded by John Major's unexpected victory in April 1992. The Conservative Party, with around 42 per cent of the electorate seemed to have a lock on the electoral system. The ability of being able to mobilise every four to five years a coalition big enough to win an apparently unbreakable majority of seats seemed to offer the prospect of of an indefinite sequence of Conservative governments. Some observers wrote off Labour after the breakaway of the SDP, a mood strengthened by Mrs Thatcher's landslide victory in 1983. Labour edged ahead of the new Liberal/SDP Alliance by a mere two per cent and some analysts (but not Mrs Thatcher herself) freely doubted whether Labour could survive as the second party. What saved Labour was what gave Mrs Thatcher majorities of a hundred and upwards on just over two-fifths of the vote – the electoral system. With fewer than 28 per cent of the votes cast, Labour in 1983 still returned 209 MPs – against the Alliance's 23. The 119 deposits that Labour lost, prove ironically to be its salvation. Its vote distribution ensured it a commanding lead over the third party in the House of Commons.

If Labour was saved by the electoral system, its new leaders' capacity to learn and to break with the soft left rhetoric of the past ranked high amongst the other reasons for the party's deliverance. The grounds for pessimism as to whether it had a future as a party of government remained. The unpopularity of the Conservative government might be demonstrated by by-elections and opinion polls, but the Conservatives had shown a continuing capacity to recover from seemingly hopeless mid-term situations – to rise again to the familiar rating of 42 per cent. Control over the timing of general elections, coupled with the ability to influence the course of the economy, gave the incumbent government a weapon of proven efficacy.

Butler and Stokes, in the early 1970s, discerned two contrasting long-term developments within the electorate.[33] The gradual decline of working-class occupations spelt for the future a steady erosion of habitual Labour support; but the distribution of party allegiance among different generations, and the effects of what these authors called 'selective death' suggested a gradual long-term swing to Labour. During its turmoil in the late 1970s and early 1980s the party threw away the latter advantage. In the meantime, a shadow loomed over Labour's calculations as long-term economic change, leading to the decline of those occupations that had hitherto heavily favoured Labour, was taking its toll. The proportion of the workforce in professional and managerial occupations, as Denver shows, almost trebled in the 25 years following 1966; this shift, coupled with the decline of industries such as shipbuilding and coal-mining, implied a swing

of 4.5 per cent from Labour to the Conservatives between 1964 and 1991.
Against such a background, the Conservative lead of 7.6 per cent in the
1992 election seemed almost predictable.

In the event neither the change in the occupational make-up of the
electorate, nor the course of the economy saved the Conservatives, though
both factors may have helped the party to limit the damage. The Labour lead
of 13 per cent was, after all, considerably below the lead shown at the start
of, and for most of, the campaign. The figures Denver cites on the
electorate's perceptions of economic improvement show that nearly a third
of the voters thought that the economy was weaker in 1997 than in 1992;
how far this reflected a genuine difference of economic experience between
this third, and the two-thirds who thought that the economy had grown
stronger or stayed the same, or simply a tendency in those who had voted
Labour to align their perceptions of the economy with their vote, is
impossible to say at present.

What allowed Labour to win, despite structural economic change and
rising prosperity was a marked increase in volatility, partly reflected in a
quantum leap in Labour's middle-class support. This raises the question of
whether the massive Labour lead was a 'one-off' feature or whether it
heralds a lasting realignment or dealignment of electoral preferences. If the
results do portend either re-alignment or de-alignment, the terms of party
debate will be profoundly changed. If the latter, New Labour, to consolidate
its gains, may have to extend its middle-England appeal even more
drastically than it did in 1997.

For the Conservatives, the most ominous finding lies in the way different
age-groups cast their vote. Only the elderly recorded a Conservative
plurality; even among the middle-aged (aged 45–64) Labour led by 10 per
cent, while among the youngest, the under-30s, Labour was ahead by 35 per
cent. The under-30s, will on average live for a good deal longer than the
over 64s. The Conservatives must hope that 'easy come, easy go', will
epitomise the responses of younger electors.

For the Conservatives, another bodeful feature of the electoral landscape
lies in the distribution of party support across the constituencies. In the past,
the great bugbear of the Liberals and their successors was their remarkably
even level of support across the country. Winning, say, 20 per cent in almost
every constituency, they could elect only a handful of MPs. In recent years
the Liberal Democrats have seen a welcome change in their pattern of
support, with greater variation from constituency to constituency. Liberal
Democrat support remains more equally spread than that of either of the
other two parties, but there are signs of improvement for the party. The
standard deviation of the vote for the Liberal Democrats has now risen to

10.9; the Conservatives, in contrast, stand not much higher at 12.2 – compared with Labour's 17.9.[34] A movement of five per cent of the poll away from the Conservatives, with the votes split equally between Labour and the Liberal Democrats, would leave the Conservatives with only 117 MPs. Double the fall in votes, on the same assumption of equal gains by the other two parties, and the Conservatives would find themselves reduced to 62 MPs.[35]

Nightmare scenarios of this kind may not survive the light of day. The probability is that the Conservatives will be the chief beneficiaries of an anti-government swing at the next election and will maintain their position as at least the second party. However, the possibility of a disaster of the kind indicated (which falls well short of Canadian meltdown) can no longer be ruled out; the odds may still be markedly against such a happening,but it cannot now be ruled out as unthinkable.

ELECTORAL REFORM

The British electoral system exacts severe penalties from parties with evenly spread support whose votes fall below a certain threshold. The chances of disaster are now big enough to make prudent Conservatives think of proportional representation (PR) as a timely insurance.

The question of the electoral system is now firmly on the political agenda. The introduction of PR for the European elections, the elections to the Scottish assembly and the Welsh assembly, offer the often lonely crusaders for electoral reform their first official recognition in over 60 years. In examining the responses of the parties to the proposed electoral system for the devolved Scottish assembly, William Miller distinguishes between paradoxical form and substantive self-interest, and argues that there is no inconsistency in proposing a proportional electoral formula for one level of government and another system for a different level.

Beyond the changes advocated for the elections to the European Parliament and the new devolved assemblies there lies Labour's manifesto commitment to a referendum on 'a proportional alternative to the first-past-the-post system, FPTP.[36] Mr Blair is reputed to be sceptical of PR and to favour either the alternative vote (AV) or the supplementary vote, the variant of AV espoused by Labour's Plant Commission. Neither of these methods can be properly be described as proportional, and it is hard to see how a referendum on either proposal could be regarded as fulfilling the manifesto pledge.

In a brief discussion in their book, *The Blair Revolution*, Peter Mandelson and Roger Liddle recommend AV as the 'electoral reform best suited to tackle the remaining unfairness in the British voting system'.[37] The grounds given fluctuate between the complacent and the banal.

Apostles for AV, enthusiasts for PR and champions of FPTP alike suffer from selective inattention. The crucial defect in the argument when the PR case rests essentially on arithmetic fairness is that the defensibility of any particular threshold, be it express,as under list systems, or implicit, as under the STV, is never addressed. Why is it regarded as outrageous that a minority of ten per cent, but acceptable that a minority of nine per cent, has no representation? Parliament must represent all the main trends of opinion, it is affirmed. But at what point does a main trend cease to be a main trend, and become an irrelevant splinter?

Supporters of FPTP rest their indictment of PR, the need for strong government apart, on two alleged weaknesses. One of these is a general charge, the other already on the edge of redundancy as a result of changes discussed in this book. PR, it is said, gives excessive power to small minorities; it also makes it impossible for the elector to vote for or against a government.

Let us take the argument about undue power being accorded to small minorities, leaving aside the obvious counter-charge that FPTP gives total power to apparently larger minorities. First, under FPTP it is impossible, because of the practice of tactical voting, to tell whether a small party is a minority or not. Second, there is nothing odd in the balance of power being held by a minority; in a group divided 51 to 49, two members will hold the balance of power and can turn a majority into a minority. This is a normal property of any divided assembly, not a special characteristic of PR. Attention would be better focused on devices to ensure that any section of members holding the balance of power uses its weight responsibly, for example, a provision for the constructive vote of no confidence. Third, why is it always assumed that the balance of power is held by one of the smaller parties? If a small party sets its price too high, the bigger parties can come together and exclude the third party. This is no empty contingency. It happened in the German Federal Republic in 1966, with the formation of the grand coalition of Christian Democrats and Social Democrats, and has happened in Israel. To hear some people talk, one wonders why politicians under proportional systems actually want to become the largest party, if the smallest party has as much power as is alleged.

Finally, of course, giving power to the smallest minority is exactly what FPTP does; we do not easily recognise it as such, because it gives this power to a small *electoral* minority – those who change their vote from one election to the next. It is those who move from Conservative to Labour, or vice versa, not the millions of faithful partisans, who determine the shape of a new government. The present electoral formula does more than give the balance of power to this minority of voters. It confers excessive power on them because of the exaggerative property of FPTP. It converts a small movement of votes into a bigger turnover of seats. It is not obvious why this should be less obnoxious than say, giving the Liberal Democrats a casting vote to decide the complexion of the next administration. (It is, of course, true that the outcome of a British general election is affected not only by straight switches between the two largest parties, but also by abstention and movement from a major party to a minor party, but this does not alter the basic argument.)

This property of FPTP also bears on the claim that the system helps provide decisive government. It may be that Mrs Thatcher's style of government could never have come to pass under PR; but it is also true that the consensus governments of 1950 to 1979, which allegedly fudged the issues and shirked hard choices, were elected under FPTP. Ministers were aware that a whisper in terms of votes lost would be magnified into a parliamentary roar of forfeited seats. Mrs Thatcher herself, for example, evaded dealing with the issue of mortgage tax relief.[38]

The second argument is that FPTP makes it possible for electors to vote directly for or against a government, whilst PR transfers the choice of government to the political brokers meeting behind closed doors. Whatever merit this view once had has been demolished by the changes in the mode of choosing the Conservative leader. The voters in 1987 chose a government headed by Mrs Thatcher that pledged to bring in the poll tax. In 1990, members of the majority party dismissed Mrs Thatcher from office and appointed a successor whose government included her chief opponent within the party. That government then abolished the poll tax. According to the quaint view which endows both the manifesto and the victorious party leader with a hallowed legitimacy, these events were 'a bad thing'. Had John Redwood succeeded in his challenge to John Major, a leader committed to the Maastricht Treaty which had been endorsed by the British public in 1992, the post of prime minister would have been filled by an opponent of the treaty.

That electoral reform might clearly be shown to be in the interest of the Conservative Party within the 'foreseeable future' may still, on balance, be unlikely. To adapt a phrase of the Duke of Wellington: there is the devil of a lot of ruin in a party. Such a vicissitude, though, should not come as a total surprise. The changes that have taken place in British politics in the last two decades have been extensive and often far-reaching. There is nothing like losing, and losing on a grand scale, to raise doubt about the fairness of the rules – except, of course, losing again and again.

The landscape of British politics has changed comprehensively within the last 20 years, not least because of the gifts and character of one remarkable woman. In a country where political leadership has for so long been a masculine prerogative, this is the biggest paradox of all.

NOTES

Acknowledgement: I wish to thank Peter Jones, Vincent Wright and Rod Hague most warmly for their advice and help in the writing of this introduction.

1. Washington Irving, *Rip van Winkle* [1820] (London: William Heinemann 1905) pp.41–3.
2. Labour Election Manifesto, Feb. 1974 and Conservative Election Manifesto Oct. 1974, in F.W.S.Craig.(ed.) *British General Election Manifestos 1900–1974* (London: Macmillan 1975) pp.400, 451.
3. R.T. McKenzie, British Political Parties (London: Heinemann 1955 and 1963).
4. R. Landes, *The Canadian Polity: A Comparative Introduction*, 4th Ed. (Scarborough, ONT: Prentice-Hall Canada 1995) p.435.
5. W. Bagehot, *The English Constitution* (Glasgow: Fontana/Collins 1963) pp.274–6.
6. *Observer*, 2 Nov. 1997.
7. Paul Webb, 'Attitudinal Clustering within British Parliamentary Elites: Pattern of Intra-Party and Cross-Party Alignment', *West European Politics* 20/4 (Oct. 1997) pp.89–110. Dr Webb's figures were derived from the database developed from the British Candidate Survey. See P. Norris and J. Lovenduski, *Political Recruitment* (Cambridge: CUP 1995).
8. Earl of Kilmuir, *Political Adventure* (London: Weidenfeld 1964) p.324.
9. McKenzie (note 3) pp.41, 46.
10. S. White, *Gorbachev and After* (Cambridge: CUP 1992) p.7.
11. Margaret Thatcher, *The Downing Street Years* (London: HarperCollins 1993) p.834.
12. Charles Wolfe, 'The Burial of Sir John Moore after Corunna', *Newry Telegraph*, 19 April 1817 from Jon Stallworthy (ed.) *The Oxford Book of War Poetry* (Oxford/NY: OUP 1984).
13. Anthony King, 'The Rise of the Career Politician in Britain – and Its Consequences', *British Journal of Political Science* 11/3 (July 1981) pp.249–85.
14. Ibid. pp.279–80.
15. See Alfred Gollin, *Balfour's Burden* (London: Anthony Blond 1965) pp.224–6. Note the extract from Sir Henry Page-Croft's autobiography *My Life of Strife* (London: Hutchinson 1948) pp.42–4. See also Peter Fraser, *Joseph Chamberlain* (London: Cassell 1966) p.285.
16. D. Baker, A. Gamble and S. Ludlam, 'Conservative Splits and European Integration', *Political Quarterly* 64/4 (Oct.–Dec. 1993) pp.420–34.
17. J.R. Vincent, *Pollbooks: How Victorians Voted* (Cambridge: CUP 1967) pp.7, 44–50.
18. M. Barker, *Gladstone and Radicalism* (Hassocks, Sussex: Harvester 1975) p.174.
19. M. Thatcher (note 11) p.753.
20. Anthony Heath, Roger Jowell and John Curtice, *How Britain Votes* (Oxford: Pergamon 1985) pp.16–20.
21. Dennis Kavanagh, *Thatcherism and British Politics*, 2nd ed. (Oxford: OUP 1990) p.291.
22. Thatcher (note 11) p.13.
23. Peter Richards, *Parliament and Conscience* (London: Allen and Unwin 1970) pp.195–6.
24. H.C. Debates, Vol.625 cols. 1453–514 (29 June 1960) Division No.126.
25. Lord Hailsham, *The Dilemma of Democracy* (London: Collins 1978).
26. R.W.K. Hinton, 'The Prime Minister as an Elected Monarch', *Parliamentary Affairs* 12/3 (1959) pp.297–303.
27. Philip Norton, *Dissension in the House of Commons 1945–1974* (London: Macmillan 1975); and *Dissension in the House of Commons 1974–79* (Oxford: Clarendon Press 1980).
28. L.W. Sumner, *The Moral Foundation of Rights* (Oxford: Clarendon Press 1987) p.1.

29. W.L. Miller, A.M. Timpson and M. Lessnoff, 'Opinions: Public Opposition to Parliamentary Sovereignty', in W.L. Miller (ed.) *Alternatives to Freedom: Arguments and Opinions* (London: Longman 1995) pp.31–44.
30. John Morley, *Life of Gladstone, Vol.III* (London: Macmillan 1903) p.497.
31. Edmund Burke, Speech on Economical Reform', in F. O'Gorman, *British Conservatism* (London: Longman 1986) p.89.
32. H.J. Laski, *A Grammar of Politics*, 4th ed. (London: Allen and Unwin 1938) p.45.
33. D. Butler and D. Stokes, *Political Change in Britain*, 2nd ed. (London: Macmillan 1974) Chs.5, 9–11.
34. H. Berrington with R. Hague, 'The Liberal Democrat Campaign', and P. Norris 'Anatomy of a Labour Landslide' *Parliamentary Affairs* 50/4 (Oct. 1997) pp.567–8.
35. Calculated by R. Hague from P. Norris (Compiler) General Election 1997 Database.
36. Labour Party 'General Election Manifesto 1997', in *The Times Guide to the House of Commons May 1997* (London: Times Books 1997) p.326.
37. Peter Mandelson and Roger Liddle, *The Blair Revolution* (London: Faber 1996) p.208.
38. For Mrs Thatcher's hostility to the abolition of mortgage tax relief and similar measures, see Nigel Lawson, *The View from No. 11* (London: Bantam Press 1992) pp.11, 362, 367, 686, 821.

Power in the Parties:
R. T. McKenzie and After

DENNIS KAVANAGH

One of the most paradoxical developments in British parties has been the shift in the contrasting internal relations of the Labour and Conservative parties. Debate over the distribution power, fired by the sociological insights of Robert Michels in pre-1914 Germany, has been a recurring theme of the literature on parties. In Britain the seminal text remains R. T. McKenzie's *British Political Parties*.[1] At the time he wrote his thesis (1955, 1963) – namely, that the distribution of power in the Conservative and Labour parties was very similar, notably in the domination of the parliamentary leadership over other sections – was highly controversial. He asserted: 'In this fundamental respect [the dominance of the party leadership] the distribution of power within the two major parties is the same' (p.582, 1st edition; p.635, 2nd edition). In the second edition he acknowledged that, for the sake of emphasis, he might have glossed over certain differences between the parties, but concluded: 'The distribution of power, as between the Leader and his front-bench colleagues, the parliamentary colleagues, the parliamentary parties mass organisations and professional machines will be [and is] fundamentally similar' (p.ix, 2nd edition).

Such a claim ran counter to the two parties' portraits of themselves and each other. Controversy was largely generated by McKenzie's analysis of Labour and his debunking of the party conference's claims to sovereignty. He had challenged the party's fundamental "myth" by arguing that the idea of intra-party democracy was (a) incompatible with the British constitution and (b) abandoned when it suited the party leaders. A large body of historians of the Labour party and leftist political scientists challenged him on grounds of historical accuracy and political values. His claims about the Conservative Party went largely unnoticed for they did not challenge the party's portrait of itself or that offered by its political opponents. Moreover, the party was little studied at the time and few were equipped to take issue with him.

McKenzie claimed that there were two major causes of leadership dominance in the parties. One was the so-called iron law of oligarchy, borrowed from Robert Michels, who had argued that control of party leaders by the mass membership was impossible. Once a party grew beyond a certain size, the emergence of oligarchy was a response to the need for organisation and specialisation, the rise of expert administrators and leaders, and the deference and loyalty which members paid to 'their' leaders. Most party members acknowledged the need for leadership and that this in turn required some autonomy for the leader(s) and restrictions on their own influence if the party were to function effectively. This logic would apply even to a party which was self-consciously anti-élitist.

But McKenzie also argued that specific features of the British political system reinforced oligarchy. The system of cabinet government, collective Cabinet responsibility to the House of Commons, and the (supposed) independence of MPs in a sovereign parliament allowed no place for a party conference to instruct MPs or the cabinet.[2] The British constitution, an amalgam of conventions, historical precedents and culture, was at odds with the constitution of the Labour party and the authority it granted party conference.

British Political Parties belongs to an élitist school of British political science, a tradition stretching from Bagehot to L. S. Amery.[3] The book is a thesis of centralised political power and independent parliamentary leadership and one in which the extra-parliamentary party is a handmaiden to the leadership. It was also part of the dominant theory of stable democracy in the 1950s and 1960s among Anglo-American political scientists. The British system, characterised by a subtle mixture of popular participation and deference to the political élites, was widely regarded as an exemplar of stable democracy.[4]

If the McKenzie thesis quickly became the orthodox view of how the parties operated, it is interesting to consider its historiography. Lewis Minkin's magisterial study of the Labour conference showed in great detail how that body's authority had declined from the late 1950s.[5] When there was a clash between conference and the Parliamentary Labour Party (PLP) leadership, the latter usually triumphed and the idea of party democracy was a casualty. In the late 1970s and early 1980s however, the pattern was reversed. Proponents of party democracy agreed with McKenzie's empirical analysis but rejected his normative approach. They concluded that reforms were necessary to make a reality of party democracy and close the gap between myth and practice. They were so successful that by the time McKenzie died in 1982 his thesis appeared to be extinct, for Labour at least. Since then, under Kinnock and Blair, the élitist model has been restored.

The central paradox, which this essay tries to explain and which runs contrary to expectations, is that over the past decade the Conservative Party, so long in government, has steadily undermined its leadership, while the Labour Party, so long out of office, has become more élitist. McKenzie tried to show that, originally, leaders were strengthened *vis-à-vis* the party outside parliament, in part because they were in government. Yet, for Labour a succession of election defeats actually resulted in an increase of the leader's authority, and conference and the National Executive became more compliant. At the same time Conservative extra-parliamentary bodies have become more assertive and willing to challenge the leadership. As Labour has become more 'old' Conservative, so the Conservatives have become more 'old' Labour. To understand Labour's evolution, Anthony Downs'[6] perspective on parties as vote-seeking organisations is a good guide to the rehabilitation of McKenzie's thesis. Just as a businessman tries to increase sales to stay in business, Labour's behaviour has been conditioned by the search for votes.

CONSERVATIVES

McKenzie certainly did justice to the tensions in the Conservative Party and the threats posed to the leadership over the first half of the century. Balfour was frequently challenged on tariffs and later on the powers of the House of Lords between 1903 and 1911, Austin Chamberlain was obliged to resign in 1922, and Baldwin was often under pressure from the party over India. Outside Parliament, National Union and the party conference became important when they reflected or exploited differences within the parliamentary élite. But in spite of the Conservative in-fighting, McKenzie had no doubt that at the end of the day the leadership in parliament was able to assert its primacy. For him, the essential pattern of conflict was between the leaders and the MPs.

In exercising authority, Conservative leaders have usually had two specific advantages over their Labour counterparts. One, already noted, is that they were so often in government. From 1922 every party leader (apart from Bonar Law, Heath, Thatcher and Hague) first assumed the party leadership when the party was in government. The premiership conferred patronage and status and these in turn attracted deference from MPs and the grass-roots. The leader's calls for support were appeals to back not just the party but also His or Her Majesty's Government. Indeed, until 1965 seven of the nine party leaders this century were chosen by the monarch, following what were termed "customary processes of consultation", rather than competitive election. It was assumed that the leader would remain in office until he died or retired, voluntarily or involuntarily. A second and more subtle reason was the high social status of many in the party hierarchy;

this could encourage every party supporter by what McKenzie called 'subtle considerations of social deference towards their leading parliamentarians'.[7] Voters deferred to Conservative MPs of breeding and within the constituency associations local grandees could exploit their superior social status. Social power buttressed political power.[8]

Yet, in the post-war period, the Conservative party has in several respects been affected by a 'contagion of the left'. After 1945 it restricted the amounts of money which candidates could subscribe to the local party. This opened the way for men and women of modest means to gain selection. In 1965 it moved to electing the party leader by Conservative MPs, in the event of a vacancy. As of 1989, however, this system had been used only twice (in 1965 and 1975) and on both occasions the party was in opposition. In 1975 the rules were amended to provide for an annual election and, therefore, an opportunity for MPs to oust the leader. The 1975 rules provided that on the first ballot a successful candidate required an overall majority, plus a lead of 15 per cent of the eligible voters (not just of those voting, as in 1965) over the nearest rival. Although in theory Conservative MPs could vote out a prime minister, this was regarded as unthinkable; after all, no Labour prime minister had ever been challenged in the annual leadership election. In 1989, however, a precedent was created when Mrs Thatcher was challenged by a relatively unknown backbencher. The following year she faced a more formidable candidate, in the form of Michael Heseltine, and the contest resulted in her exit. In 1995 John Major had to overcome a challenge by John Redwood.

The holding of an annual leadership election and the possibility of a contested ballot is now a significant limitation on a Conservative Party leader or prime minister. (A Labour leader, see below, is better protected.) The previous three Conservative leaders before Major were each forced to stand down, two as a result of a leadership election and over six years (1989–95) there were three leadership challenges. Each year there is the possibility of rival teams of supporters and programmes emerging as candidates strive to build up support. Candidates may include fellow cabinet or shadow cabinet colleagues who wish to place a marker for the future. Moreover, having used the election system to oust two party leaders (in 1975 and 1990) and destabilise a third (John Major), the annual election system is potentially a significant anti-leadership force and one which enhances the power of backbenchers. James Douglas, a former Conservative official who played an important role in drawing up the 1965 and 1975 rules, has pointed out that few of his colleagues at the time foresaw the effect of the elections and campaigning by rivals in hardening in the party divisions into factions: 'the experience of campaigning together

for a particular candidate and the animosities generated by the campaigns
… gave each "tendency" a persistence and sense of permanence'.[9]

The relative frequency of elections for the leadership has triggered a
growing demand for the members to participate, just as members do in the
Labour and Liberal Democrat parties. This gained added strength in 1997
when John Major stepped down. Restricting the vote to the 164 available
MPs meant that three-quarters of local associations, including Scotland and
Wales which did not return a single Conservative MP, had no formal voice.
At present the Conservatives have a leadership election system which in a
sense is both élitist (confined to MPs) and anti-élitist (because of the ease
with which the incumbent can be challenged). In introducing elections
among MPs for a vacancy in 1965 and then allowing the leader to be
challenged in 1975 the party shifted towards the Labour system. Labour
moved the goal posts again in 1981 by extending the suffrage to party
members. At the time, the removal of this power from MPs exclusively was
opposed by many commentators, who regarded it as a formula for control
by left-wing activists. These fears have not been borne out. The
Conservatives are now under pressure to emulate this last step.

Richard Kelly has done much to revise our interpretation of the role of
the Conservative Party conference.[10] For long it was regarded as merely a
'dignified' part of the political process, largely because our understanding
has been dominated by the Labour conference, with its open system of
vigorous debate, card votes, compositing resolutions, block votes and
making of decisions capable of sharpening party policy. Kelly's point is that
the Conservative representatives' agreement with and support for the
leadership was often less a mark of deference by the former than a
consequence of the latter anticipating or responding to concerns that had
been articulated in the series of conferences that had been held in the
months before the annual conference. Kelly argues that this is a 'hidden'
Conservative conference system in which the leadership is expected to
demonstrate that it has taken on board the concerns of members; dissent
breaks out when it fails to meet those concerns. Conference influence was
clearly seen in 1987 when pressure from the floor forced what turned out to
be a disastrous decision to introduce the poll tax in one fell swoop, rather
than gradually over four years as had been planned. The bitter European
debate in the 1992 conference demonstrated a lack of deference among
Eurosceptics, and 'the floor was no longer content with subtle, coded
criticism'.[11] Away from the conference floor the growing number of fringe
meetings provides ample opportunities for the different factions and
leadership aspirants of the party to disport themselves.

But here is another paradox. At a time when Labour's conference has become increasingly stage-managed and more of a rally to support the leadership, so the Conservative conference has come to resemble the old Labour model of providing abrasive debate and threatening the serenity of the leadership. In the 1990s constituency representatives at party conferences have been encouraged by the example of dissident MPs and former senior party figures. The conferences have increasingly mirrored and exacerbated the divisions among MPs, particularly over Europe.

LABOUR

Historically, there have been two patterns of power relationships in the Labour Party. In opposition, the extra-parliamentary bodies, particularly the National Executive Committee (NEC) and the party conference, were assertive and influential, and the party leader and members of the shadow cabinet were more beholden to these bodies. But in government the leaders were usually more orientated to national responsibilities than to conference and NEC resolutions and entreaties. In opposition leaders could agree with these bodies but as ministers they had to take actions which sometimes led to conflict with both. Lewis Minkin has shown how after 1964 and again after 1974 Labour ministers increasingly defied conference – over prices and incomes policies, membership of the EEC (as it then was), economic strategy and public-spending programmes. Conference defeated the platform only twice between 1945 and 1966, but did so on 32 occasions between 1970 and 1979. Harold Wilson and then James Callaghan lectured conference that Labour was now a party of government and its duty was to govern. The NEC, however, increasingly presented itself as an alternative voice of the party and conference.

So far, Labour ministers had acted out the McKenzie analysis of the party. When the party leadership was in government and disagreed with conference, the theory of inner-party democracy was questioned. Activists concluded that winning policy battles when Labour was in opposition was pointless if the leadership abandoned or diluted them when in government; they should, therefore,concentrate on reforming the party. They campaigned for reforms to make the PLP more accountable to conference and MPs more accountable to constituency activists. The goal was to make the Labour Party safe for socialism by closing the gap between the myth of inner-party democracy and the reality which McKenzie had analysed.

The success of the activists produced what, after 1979, was virtually a new Labour Party, to anticipate the terminology of Tony Blair. Two key 'democratising' reforms – mandatory reselection of MPs within the lifetime of parliament and election of the party leader by an electoral college of party members, with most of the vote going to trade unions and constituency parties – were achieved. A third reform, giving the NEC control of the manifesto, was only narrowly voted down. The leadership election scheme adopted in 1981 gave MPs only 30 per cent of the vote compared to the trade unions' 40 per cent and the constituency parties' 30 per cent.

The changes in the power structure also had consequences for academic debate. McKenzie had been refuted, 'his more controversial claim about the similarity between Labour and Conservative parties in their internal power relationships is no longer true'.[12] Lewis Minkin revised his earlier conclusion about the decline of party democracy and suggested: 'The party's power relations have proved to be much more dynamic than was once supposed, capable of being reshaped from below as well as from above, susceptible to periodic change between the political and trade union leadership.'[13]

Several factors combined to undermine the parliamentary leadership's domination. One was the widespread sense of disappointment with the records of the Wilson and Callaghan governments, coupled with a belief that in defying conference they had 'betrayed' the values and ideas of the movement. Moreover, the governments had also lost the 1970 and 1979 general elections. Revisionists had claimed that their policies would promote economic growth, and win elections. The failure of the revisionists to provide ideological comfort or win elections enabled the left to gain a better hearing.

Another factor was the changing role of the trade unions. For much of its history the parliamentary leadership had relied on the major trade unions to act as a praetorian guard and head off the radical demands of constituency parties and left-wing MPs at conference.[14] Co-operation between the party leaders and the big unions broke down in the late 1960s, however, as the unions were radicalised by Labour ministers' imposition of incomes policies and attempts to legislate in the field of industrial relations. Free collective bargaining was a key issue for the trade unions, a *sine qua non* of their existence. Union activists became more left wing and union leaders were no longer able to deliver the vote of their executives for the party leadership. The activists were increasingly ideological, middle class, employed in the public sector and assertive, not least in making demands for greater spending on public services.

Finally, the PLP was also shifting to the left, partly as a net result of the retirements and replacements of MPs and partly as a result of the flight of many disillusioned right-wingers in 1981 to the SDP.[15]

The new arrangements were closely related to policy battles between the right and the left. Changing the rules to party democracy had far-reaching political consequences, in the short run at least. If the old informal rules of the game were tailor-made for the revisionists, or the right wing of the party, the new rules were tailor-made for the left. Although the reforms were strongly criticised at the time by political commentators they only brought the practices of the Labour party into line with what happened in many other countries. Indeed, in 1997 Conservative reformers are considering Labour's leadership election rules as a model for their party.

The main objection to the rule changes from most commentators and many party leaders at the time was that they were promoted by the left and the left gained from them in policy terms. The party fought the 1983 general election on its most left-wing manifesto since 1935. Particularly instructive was the emergence of the manifesto. The Clause V meeting, in which the NEC and shadow cabinet agree the contents of the manifesto, saw the draft document pushed through with little discussion or amendment. In the shortest Clause V meeting ever, shadow ministers had little or no impact on a conference (and, therefore, left-wing inspired) document. The right was weakened anyway because several of its leading figures had left to form the Social Democrat Party. The changes in Labour's constitution, which they feared would result in a permanent diminution of their wing's influence, was the final stage in their alienation from the party

The 1983 election defeat, however, was a watershed. The newly 'democratic' party gained only 27.6 per cent share, its lowest figure since it was formed as a national party in 1918. Since then power in the Labour Party and policy have shifted in one direction only, as successive reforms of the institutions and a change in the ethos of the party have empowered the leader and his office, and many left-wing policies have been abandoned. The steady decline in size of the groups which provided the party's core vote – the working class, trade union members and council tenants – and survey evidence of the unpopularity of many policies, forced the shift. The leadership's determination to become more responsive to the electorate at large, rather than to the activists, clearly challenged the inner-party-democracy model. A theme of the academic literature is that many rank-and-file party activists are more radical than party leaders, on the one hand, or the voters, on the other.

This produces a curvilinear relationship between radicalism and the individual's location in the hierarchy of the party organisation.[16] It suggests that Labour activists will be to the left ideologically of both the party's leaders and its voters.

The party had to find new sources of support. Rebuilding and reforming the party were necessary if Labour was: (a) to abandon electorally unpopular left wing policies and (b) to make its image more attractive to the electorate.

Neil Kinnock explained in 1994 how difficult it had been for him to achieve the policy and organisational changes while he was leading the Labour Party in the mid-1980s.[17] At the outset he lacked an assured majority on the NEC. He supported the proposals for one-member, one-vote (OMOV) for the selection and re-selection of candidates but saw them turned down at the 1984 conference, largely by the trade unions whose power would have been diluted as a result. Another comprehensive general election defeat in 1987 proved a spur for Kinnock to go further. He commissioned a major review of party policy, which was dominated by the parliamentary leadership at the expense of the NEC. The review detached the party from many unpopular left-wing commitments, particularly on public ownership, re-nationalisation of recently privatised utilities, industrial relations and unilateral nuclear disarmament. Opposition to the modernisation of the party structure and the policy changes came from the left.

In 1988 the NEC, by this time effectively under Kinnock's sway, took power to intervene in the selection of by-election candidates, a power which was usually employed to exclude candidates on the left. The principle of OMOV was extended for the elections to the constituency section to the NEC; hitherto this choice had been left to the activists. The effect was seen in the election to the NEC of MPs more supportive of the leadership. By 1990 mandatory re-selection of MPs was effectively overturned when conference agreed that re-selection contests would only take place if a ballot of Constituency Labour Party (CLP) members demanded one. In 1988 steps were also taken to strengthen a Labour prime minister against a formal leadership challenge, and to boost the influence of Labour MPs in the election process. A leadership candidate now has to be nominated by a minimum of 20 per cent of MPs, compared with the previous five per cent and, when the party is in office, two-thirds of the conference have to approve a challenge. This makes a Labour leader and prime minister more secure than his Conservative counterpart.

The influence of conference over policy was downgraded by the creation of the Policy Forum, consisting of several commissions on which the parliamentary leadership played the initiating role. Since 1991 shadow-cabinet members have been allowed to address conference from the platform; before they could only do so if they were also members of the NEC. In the 1987 and 1992 general elections a Shadow Communications Agency, drawn largely from members of the public relations industry, was established and given a key role in developing campaign strategy and marketing the party. Critics of the 1992 campaign complained that the body operated with a good deal of independence from the NEC and other party bodies.

Since Kinnock's resignation after the 1992 general election there has been no let-up in attempts to reverse the effects of the reforms of the early 1980s. At the 1993 conference, John Smith fought for and achieved OMOV, effectively breaking the block vote of the trade unions. The trade-union share of the electoral college to elect the party leader and deputy leader was reduced from 40 per cent to a third, the same share as MPs and constituency parties. The union share of the conference vote was further reduced to 50 per cent in 1994 when constituency membership reached 300,000.

The Policy Forum and the shadow cabinet and the leader rather than the party conference now play the major role in policy formation. According to party literature the forum debates should be conducted in a 'purely advisory' manner, avoid votes and rely on senior party figures to 'sense the opinions' of speakers. The forums are designed to iron out disagreements before they get to conference. Under Blair the party conference has emerged more than ever as a rally to support the leader and his vision of New Labour.

Kinnock, Smith and Blair have also used ballots to mobilise 'ordinary' party members to outflank the activist democracy beloved of the left. John Smith was elected party leader in 1992, with over 90 per cent of the vote of CLPs, most of which held ballots of members. The Labour membership in June 1995 approved the scrapping of Clause IV which committed the party to public ownership, and the draft election manifesto in October 1996. In both cases virtually all local parties held ballots of members. Essentially, however, these were changes offered on an all or nothing basis and the resources of the party leadership were thrown behind the desired outcome. Members had little choice but to approve the entire draft manifesto, faced as they were with a plebiscitary style of leadership.

The party has increasingly used opinion polls, advertising and public relations to draw up policies and project themes which appeal to target voters, often voters who are 'weak' Labour or potential converts from the Conservative Party. In the 1997 general election the leadership concentrated its efforts on providing reassurance for these key voters. It is the opposite of the activist democracy the left fought for in the 1980s.

The new leadership-centred party structure is designed to continue if and when Labour is in government. In January 1997 the NEC approved proposals to transform itself and the annual conference. The document *Labour into Power: A Framework for Partnership* proposes a broadening of the NEC's membership to include, for example, representatives of the cabinet and PLP as of right, giving oversight of policy to a joint cabinet-NEC committee chaired by the party leader, and reforming conference to provide opportunity for more detailed discussion about policy proposals. The NEC and conference will have a responsibility to support a Labour government. The extra-parliamentary party is to become the hand-maiden of the parliamentary leadership, as in the bourgeois model.

CONCLUSION

What might explain these major shifts in the parties' internal relationships? Are the reasons common to the Labour and Conservative parties, or is there a particular combination of issues, leaders and life-cycle in each party's role in opposition or in government? What seems to be important is less the British constitution than (a) for Labour, power politics, namely politicians' and others' calculations about how to win elections and (b) for Conservatives, the impact of fundamental issues which raise divisions about a party's identity.

In the case of Labour, the chief explanation for the big changes in structure and ethos seems clear enough – the electoral imperative, or the requirement of winning general elections in a competitive party political system. The McKenzie model has been reinstated, but for reasons which have little to do with his original analysis, namely, the parliamentary system. More important have been the reactions to the comprehensive electoral defeats which followed the exercise in inner-party democracy, calculations about the steps necessary to become electorally competitive, and the organisational and cultural consequences of new campaigning techniques. Like other modernising European socialist parties, Labour has become what Panebianco calls an 'electoral-professional' organisation.[18] The party leadership from Kinnock onwards regarded the old model Labour Party as ineffective on three grounds:

- *Campaigning.* An effective campaigning party today requires a centralised authority structure, so that it has the speed and flexibility to respond to changing circumstances. It also needs to be voter-responsive rather than activist-oriented. The institutions and procedures of 'old' Labour gave undue weight to the views of party activists who were atypical of party supporters. Party leaders have used the findings of opinion polls and focus groups as guides to public opinion, rather than the resolutions of committees dominated by activists and trade union delegates.

- *Image.* The inner-party democratic model in the 1970s and early 1980s did little to project Labour as competent, united, strongly led and 'fit to govern', in fact quite the opposite. Yet, much evidence, from surveys and from Labour politicians' personal contacts with party supporters, showed that such qualities were important to voters.

- *Television.* As television has become the primary means of political communication, so most politicians have become increasingly self-conscious or manipulative in their approach to the medium. Continuous television coverage of the annual party conference, for example, has transformed it into more of a theatrical event, requiring more stage management. The significant interaction is between leaders and party members, on one side, and voters at home in their sitting rooms. The televised conference (like election press conferences) provides a platform for the political party to appeal to the electorate and has become part of the permanent campaign. Coverage of bitter debate or shows of disunity usually damages the party's image. If the leader is shown in a confrontational setting in conference or parliament it is best presented as one in which he or she faces an unpopular minority and is seen to 'win'. The classic example of this was Neil Kinnock's attack on Militant in his 1985 conference speech.

These factors have encouraged Labour to become more of a leader-driven party, more like the old Conservative Party. Just as the battle over party reform in the early 1980s was inseparable from left–right disagreements on policy, so also adoption of new campaign methods and steps to increase the autonomy of the leadership were proof of a larger modernising project to marginalise the party's left wing and shift party policy to the middle ground.[19]

The Conservative Party has, of course, moved in a different direction, one resembling the Labour Party of the late 1970s and early 1980s. This has produced less a convergence between the two political parties than one in which they have bypassed each other on a continuum of leadership authority. Whereas lengthy exclusion of office has led to élitism in Labour, the lengthy period of office has produced the opposite reaction among Conservatives. Why has Tory electoral success – so long regarded as a key criterion of successful leadership – resulted in the undermining of the leadership?

We can reject, at the outset, an explanation which merely emphasises the replacement of Margaret Thatcher by John Major and the latter's allegedly 'weak' leadership. The forces undermining party unity and the leadership's authority were clearly apparent before 1990. Some MPs were embittered by Mrs Thatcher's downfall in 1990 but one should also recall the discontent of the pro-European wing with Mrs Thatcher in the late 1980s. As noted, the development of a more participatory and assertive conference was evident in the decade. The National Union pressed for party members to have a more formal role in leadership ballots. In the 1997 contest for the party leadership, few MPs appear to have been influenced by the 'advisory' votes of constituency chairmen, who favoured Kenneth Clarke over other candidates over all three ballots. It is now almost inevitable that the Conservative Party will follow Labour and find a way of involving members in the election of the leader.

It is, therefore, important to refer to three other broader factors which have been at work in the Conservative Party. The first is the growing habit of dissent among MPs. Philip Norton has charted the growth of back-bench dissent under the Heath government between 1970 and 1974.[20] That it grew under Mrs Thatcher might have been explained in terms of her government's large parliamentary majorities. MPs could withhold support in the division lobby, confident that they would not bring down the government. But after 1992 dissent continued under John Major, particularly over Europe, when his government had only a very narrow majority and at times withdrawal of support imperilled the life of the government itself. In November 1994 the whip was withdrawn from eight MPs who had abstained on the European Communities (Finance) Bill, relegating the government to a minority in the Commons. The severity of the step is shown by the fact that between 1905 and 1992 only five Conservative MPs had had the whip withdrawn.[21]

The second is the emergence of a key issue which divides the party. British political parties have been fairly successful at processing issues and presenting a united front to the electorate. But there have always been major issues which have presented acute problems of party management, for example, appeasement, for the Conservatives in the late 1930s and Enoch Powell's stands on immigration and membership of the European Community in the late 1960s and early 1970s. Europe has provided a major faultline in the party over the past decade. It figured prominently in Sir Geoffrey Howe's resignation from the cabinet in 1990 and the events which forced Mrs Thatcher's retirement from the leadership in 1990 as well as the earlier resignations of Heseltine, Lawson and Ridley from her cabinets. It continued to divide the party under Major.

Some of the Conservative dissenters under John Major were Thatcherites MPs who strongly support national sovereignty and free market economics.[22] Rebellions in the House of Commons division lists were largely on the passage of the Maastricht Bill and other European issues. The research team at Sheffield University has pointed out that the European divisions in the 1990s have parallels to previous Conservative splits in 1846 over the repeal of the corn laws and in 1903 over tariffs.[23] The party has long contained factions which disagreed about the importance of Britain's national sovereignty, between those who thought that Britain should stand alone and those who believed that it would have to develop close relationships with other states, largely for economic reasons. In the 1990s the integrationist push in Europe, reflected in the extension of majority voting, the social chapter, Maastricht and the single currency, has struck a sensitive nerve in the party. An increasingly integrated and internally regulated Europe offended the Thatcherite vision of the independent state and the market economy. Sovereignty was instrumental for the free marketeers, as it was for the Labour left from the late 1940s onwards. The Major cabinet's line of 'negotiate and then decide' whether Britain should enter the first stage of the European single currency in 1999 failed to satisfy the sceptics. In the 1997 general election over 200 Conservative candidates broke with the manifesto and flatly declared that they would vote against British membership of a single currency.

One can certainly argue that changes in the European project in recent years have helped to make the Conservative Party the chief home of the Eurosceptics. But Tory MPs have also changed; they are more full-time, careerist and choose para-political posts early on which will assist their entry into politics. These pre-parliamentary jobs might be, for example, in the party's organisation, in parliament or Whitehall, as a special adviser, in the media, in a think-tank, or in a company lobby or pressure group.[24] The Conservative Party may or may not have become more ideological and more divided, but it has certainly become more factionalised; witness the development in recent years of the Bruges Group, No Turning Back, Charter Movement, or Conservative Way Forward.

Finally, it is important to identify accurately where the tensions exist in the party. It is less a case of the National Union, the annual party conference or even the constituency associations challenging the party leadership, although demands for rank-and-file participation in the election of the party leader have become more vociferous. Most of the time these bodies are appealing for greater unity among MPs and support for the leadership. The dissent and the challenge to Major came primarily from MPs on the right, although at times these were backed by their constituency associations. When MPs have been de-selected – as, for example, David Ashby, Sir George Gardiner and Sir Nicholas Scott were in 1996 and 1997 – lack of local support has had more to do with the perception that the MP's public or private behaviour was damaging to the party, than pressure from the leadership. During the 1997 general election campaign, it would have suited the Conservative leadership if local parties had dropped some candidates associated with reported sleaze, but the local parties rallied behind their MPs.

Are political parties more or less central to political life in Britain, compared with 20 years ago? They clearly remain important for recruiting members to parliament and ministers in government. But in many respects political parties seem to be a declining force. Membership has clearly fallen from the 1950s and 1960s, in spite of some revival in the Labour Party since 1994. Perhaps less than two per cent of the electorate is actually a member of a political party, a lower membership–voter ratio than in many other countries. Members remain important as a source of funds, as a pool for filling local political offices, and beating the drum for the party on local platforms and writing letters to the local press. But people wishing to be active and to promote a cause seem more willing to engage in an interest group than in a political party. There is evidence also of declining activity, or what has been called a 'de-energising' of local members.[25]

Finally, campaigning political parties, like cause groups, have increasingly called upon the aid of professionals in marketing and public relations and think-tanks for policies. Political parties now employ their own experts in opinion polling, direct mail, fund-raising and advertising. The targeting of key voters in marginal seats is increasingly conducted from the centre, via direct mail or telephone canvassing, rather than by local activists. This last development gives rise to another paradox. What matters less than ever to each political party is its strong partisans.

The declining number of partisans means the parties have to reach out for other supporters if they are to forge an electoral majority. Increasingly, the target for any political party is the 'soft' vote, namely those who have recently switched to or who are thinking of defecting from it. Such voters are likely to be only weakly attached to political parties and not very interested in politics or issues, but more interested in image and style. The concentration of the political parties on these target voters is likely to produce some convergence of the parties' political communications, and an emphasis on image rather than substance.

NOTES

1. R.T. McKenzie, *British Political Parties* (London: Heinemann 1955 and 1963).
2. Although party whips instructed MPs how to vote.
3. L.S. Amery, *Thoughts on the Constitution* (London: OUP 1947).
4. See G. Almond and S. Verba, *The Civic Culture* (Princeton UP 1963).
5. *The Labour Party Conference* (Manchester UP 1980).
6. Anthony Downs, *An Economic Theory of Democracy* (NY: Harper Row 1957).
7. McKenzie (note 1) p.638.
8. S. Beer, *Modern British Politics* (London: Faber 1965).
9. 'Conservative Party Leadership Selection Process', *British Politics Newsletter* (Winter 1996) pp.6–10.
10. Richard Kelly, *Conservative Party Conferences* (Manchester UP 1989).
11. 'The Power of the Tory Conference', *Spectator*, 5 Oct. 1996, pp.21–3.

12. D. Kavanagh, 'Power in British Political Parties: Iron Law or Special Pleading?', *West European Politics* 8/1 (Jan. 1985) pp.5–22.
13. L. Minkin, *The Contentious Alliance: Trade Unions and the Labour Party* (Edinburgh UP 1991) p.366.
14. On this, see L. Minkin, ibid. and M. Harrison, *Trade Unions and the Labour Party Since 1945* (London: Allen & Unwin 1960).
15. See H. Berrington, 'The Labour Left in Parliament: Maintenance, Erosion and Renewal', in D. Kavanagh (ed.) *The Politics of the Labour Party* (London: Allen & Unwin 1982).
16. J. May, 'Opinion Structure of Political Parties: The Special Law of Curvilinear Disparity', *Political Studies* 21 (1973) pp.135–51.
17. N. Kinnock, 'Reforming the Labour Party', *Contemporary Record* 8/3 (Winter 1994) pp.535–54.
18. A. Panebianco, *Political Parties: Organisation and Power* (Cambridge: CUP 1988); R. Katz and P. Mair (eds.), *How Parties Organize* (London: Sage 1994) and D. Kavanagh, *Election Campaigning. The New Marketing of Politics* (Oxford: Blackwell 1995).
19. See E. Shaw, *The Labour Party Since 1945* (Oxford: Blackwell 1996) and R. Heffernan and M. Marqusee, *Defeat from the Jaws of Victory: Inside Neil Kinnock's Labour Party* (London: Verso 1992).
20. P. Norton, *Conservative Dissidents: Dissent within the Parliamentary Conservative Party 1970–74* (London: Temple Smith 1980).
21. P. Norton, 'Losing Whips and Votes', *British Politics Newsletter* (Winter 1995) p.306.
22. P. Cowley, 'Philip Norton's Conservative Party', ibid. (Fall 1996) pp.16–18.
23. D. Baker, A. Gamble and D. Ludlam '1846 ... 1906 ... 1996? Conservatives Splits and European Integration', *Political Quarterly* 64 (1993P pp.420–34.
24. P. Riddell, *Honest Opportunism: The Rise of the Career Politician* (London: Hamish Hamilton 1993).
25. P. Whiteley, P. Seyd and J. Richardson, *True Blues: The Politics of Conservative Party Membership* (Oxford: OUP 1994).

Europe, Thatcherism and Traditionalism: Opinion, Rebellion and the Maastricht Treaty in the Backbench Conservative Party, 1992–1994

HUGH BERRINGTON and ROD HAGUE

The Treaty of European Union signed by the governments of the member states at Maastricht in February 1992, with its provisions for the extension of majority voting, the adoption of the Social Chapter and timetable for a single currency and monetary union, marked a turning point in the development of the European Community. Most observers, whether they welcomed or opposed it, saw Maastricht as a milestone on the road to European federation. For some members of the Parliamentary Conservative Party, Maastricht was (as Margaret Thatcher put it) 'a treaty too far'.[1] The procedure of parliamentary ratification of the Treaty (despite the 'opt-out' from the Social Chapter negotiated in December 1991) ignited a long-running rebellion. In effect, the issue of European union has exposed a major faultline within the parliamentary Conservative Party,[2] which at times posed a severe threat to John Major's government. In so doing, it has given investigators an excellent opportunity to examine a range of problems: among them, the interplay of opinion within the backbench Conservative Party, the relationship between opinion and division-lobby behaviour, the link between demographic variables (age, educational background, constituency characteristics, etc.) and the pattern of attitudes towards Europe within the parliamentary Conservative Party .

THE APPROACH

We have employed several sources of data, and our techniques have included variants of roll-call analysis. For our sources we have relied on Early Day Motions (EDMs) supplemented by floor revolts and free votes in the House of Commons and lists of MPs proclaiming support for particular causes. The utility of EDMs as a source of data for tracing backbench opinion has been extensively discussed.[3]

In brief, EDMs are motions put down by backbenchers; their subject matter covers a wide range of topics, some politically controversial. While some EDMs are designed as mere demonstrations by one member or a handful of MPs, others attract many signatures. The number of EDMs has exploded in the last few decades. In the session of 1992–93, which ran for 18 months, nearly 2,600 (plus amendments) were tabled, whilst even in the normal year-long session of 1993–94, over 1,700 (plus amendments) were put down. We selected for analysis a number of EDMs tabled in these two sessions. We eventually chose 11 EDMs signed by Conservative MPs, recorded the signatures, calculated Phi intercorrelations and then scaled the motions using Principal Components Analysis. From this information we sought to construct two scales – a Maastricht scale, which essentially measured opposition to Britain's ratification of the Maastricht Treaty, and a 'wet–dry' scale, based largely on sympathy for public expenditure on overseas aid and the disabled.

There have been considerable changes in the composition of the parliamentary Conservative Party over the past 20 years. In the 1950s, the backbench party was drawn largely from élite sectors of British society, and was socially homogeneous.[4] In the 1960s and 1970s, the public school/public company hegemony began to erode, and this decline accelerated during the long period of Mrs Thatcher's premiership. Burch and Moran showed the drastic changes taking place; in the *new* intake of 1983, Etonians constituted only six per cent.[5] In policy terms, the party became ideological and programmatic, and this break with its past was paralleled by the decline of the social groups which had hitherto been the most prominent element of the parliamentary party.

The Maastricht revolt was one of the most long-lived and the most acrimonious in Conservative post-war history. Can we explain the tensions over Maastricht by the influx of new social groups in the 1970s and 1980s? Do the Maastricht rebels represent a new breed of Conservatives – or is there evidence to support the charge that the revolt reflected the bitterness of those whose ministerial career hopes had been frustrated?

A further question is how far the anxieties about European integration mirrored fears about the fate of the Thatcherite agenda in a post-Maastricht Europe. Were those who rebelled prompted by worries that Britain would be bound, against its will, by collectivist Euro-legislation?

There remains one last theme – that of law and order, or more specifically, capital punishment, together with a related issue, that of corporal punishment in schools. Were hostility to further European legislation and support for the restoration of the death penalty and school caning fed by a common source? The free votes of February 1994 and a division in January 1997 offer simple indicators of members' opinions.

This approach is exploratory. We began by seeking to distinguish three separate faultlines in the Conservative Party – the cleavage over Maastricht, the wet–dry divide and the conflict over capital punishment – to see how they related to each other, and to determine what distinctive social sources, if any, nourished the ideological groups we sought to locate. We used various kinds of indicators. EDMs put down in the two sessions of 1992–93 and 1993–94 were our main source for attitudinal indicators. Floor revolts and a free vote, derived from the division lists, and newspaper reports of deliberate abstentions (not alas recorded by Hansard), furnished *behavioural* indicators.[6] With these sources we can ask how far attitudes, at odds with those of the party leadership, were translated into parliamentary rebellion. The free vote on the death penalty, together with the whipped vote on caning in schools, affords us both *attitudinal* and *behavioural* indicators. Finally, wherever possible we have sought to validate and extend our conclusions with evidence drawn from independent published lists of members supporting the manifestoes of specific organisations, or from jointly signed letters.

The number of EDMs put down in the two sessions was huge – a total of 4,300. Diversity of subject matter matches the number of motions. EDMs range from the trivial to the profound, from partisan squibs to demonstrations of internal party divisions. We excluded all EDMs, irrespective of their content, with fewer than nine Conservative signatories. A vast number still remained, but many of these were Labour-inspired. Of the Conservative-signed EDMs we initially selected 37. We eventually chose 11 motions on grounds of (i) their substantive policy interest and (ii) signature correlation with other EDMs, having calculated Phi correlations between the motions that were of substantive interest to us. The 11 contained four EDMs central to the Maastricht dispute of which EDM 92/174 (Fresh Start) is the most famous. The other seven included one on economic policy and six calling for either increased expenditure on the disabled or on international humanitarian aid. A full list of the 11 is given in Appendix 1 (p.69).

MEASUREMENT: SOURCES AND METHOD

The Anti-Maastricht Scale

We applied Principal Components Analysis to the subset of EDMs originally selected for scrutiny. The first component was clearly an anti-Maastricht dimension. The four anti-Maastricht EDMs had loadings ranging from –0.512 to –0.408. All the other EDMs, bar two, had positive loadings;

the remaining two were weakly negative. Members were scored on this anti-Maastricht dimension according to the sum of the loadings of the motions they signed. Thus, Christopher Gill (one of the 'whipless nine') signed the four anti-Maastricht motions but no others and was therefore credited with a score of -1.818 (the sum of the loadings of the four anti-Maastricht motions). We were therefore able to rank the 155 MPs who signed at least one motion in the subset, with Sir David Knox, a strong Europhile emerging as the least anti-Maastricht member and Richard Shepherd, who like Christopher Gill signed all four anti-Maastricht EDMs, coming out as the most hostile of the Eurosceptics.

On inspection, the party divided into almost equal sections. There was a clear break in the scores between -0.263 and -0.043; for purposes of analysis we termed those with scores of -0.263 and lower anti-Maastricht and those with scores above, pro-Maastricht, though we acknowledge that among the latter enthusiasm for the Treaty varied considerably. We suspect that those we have called pro-Maastricht include some of the non-aligned, responding to the call for party loyalty.

We do not pretend to assess the absolute level of Euroscepticism or Europhilism. Scales of this kind are *ordinal*; they show the position of members relative to one another. For analytic purposes, our choice of cut-off had the advantage of dividing the party into two categories almost equal in size. Moreover, inspection of these two categories, by reference to signature of EDMs and floor revolts, showed that this cut-off point marked a substantive division within the party.

The 76 we defined as pro-Maastricht contributed only two signatures between them to the four anti-Maastricht EDMs. In the seven floor revolts we examined, they did not muster a single dissenting vote, nor even an abstention. Each member of the anti-Maastricht group signed at least one of the anti-Maastricht EDMs. Twenty of them rebelled in five or more floor revolts, and a further 26 went into the division lobbies against the government at least once. Another abstained twice.

The Capital and Corporal Punishment Votes

A debate on capital punishment, in February 1994, provoked two divisions in which, as is customary on such issues, the government allowed a free vote. One division called for the restoration of the death penalty for all kinds of murder, subject to certain safeguards, whilst the other demanded the death penalty for the murder of police and prison officers on duty. We chose the broader division on the general principle of capital punishment. Forty-eight of the backbenchers in our database voted against bringing back capital punishment whilst 78 voted for the death penalty. In the 1996–97 session the government's Education Bill gave backbenchers an opportunity to move an amendment to bring back corporal punishment in schools in specific circumstances. Ninety-one MPs in our database voted for school caning and 109 followed the whip in voting against.

The Wet–Dry Division

It is not easy to classify Conservative backbenchers according to their opinions on welfare expenditure or economic policy. Our attempt to develop a wet–dry scale, to complement our Maastricht scale, employing Principal Components Analysis and using EDMs on international humanitarian aid and the treatment of the disabled, was not successful. The difficulty may lie in the sparsity of our data, or perhaps in the very notion itself of a coherent and simple wet–dry cleavage. We considered using a single motion, Economic Policy (EcPol) tabled in October 1992, to distinguish wet and dry backbench MPs. The content of this EDM seemed essentially dry in orientation, and our interpretation was strengthened by a breakdown of the signatories, according to their response to the humanitarian aid and disablement EDMs. Nevertheless, the content of EcPol did not seem unambiguously dry and we remain chary of relying on a single motion. We report later, without laying great stress on our findings, an analysis of the demographic attributes of the welfare/international aid motions.

Total Non-Signers

Seventy-three members, out of our database of 228, failed to sign a single motion from our subset of 11 EDMs. We call these the 'total non-signers' (though we know that some of them signed motions we scrutinised but discarded); nearly half of them were parliamentary private secretaries (PPSs) and a further 20 were ex-ministers who, we assume, wished to keep their distance from the factional activities of their backbench colleagues. The number of total non-signers who fell outside these categories is very small.

In this study, we can strictly identify only the anti-Maastricht MPs, for there were no pro-Maastricht EDMs. We characterised those who did not sign *Fresh Start* or its successors as pro-Maastricht and, as we have seen, there is considerable evidence to support this judgement (see above, though as we have also said, our pro-Maastricht group probably includes some of the non-aligned). By the same token it can be argued that the total non-signers conceal a substantial body of opinion favourable to the Treaty. Thus, those total non-signers who rebelled, whether by reported abstention or an adverse vote, against the government during the bill's passage, were few in number; and 18 of the total non-signers signed the Positive European

Group's declaration in January 1995. However, for purposes of analysis, to qualify as pro-Maastricht a member had to sign at least one of the motions in our subset other than the four anti-Maastricht EDMs.

THE MAASTRICHT CONTROVERSY: BACKGROUND VARIABLES

The hypothesis we wished to explore was that the turbulence over Maastricht reflected the changes in the party's social composition which began, if slowly, in the 1960s and accelerated during the Thatcher epoch. The intake of new MPs in 1979 and after was drawn much more than hitherto from the lesser public schools and the state schools, and from those with experience in family businesses rather than large public companies. In so far as these MPs favoured a rigorous cutting of public expenditure and continuing government disengagement from the economy and welfare provision, they would see the ever closer union foreshadowed by Maastricht as a threat to these values. For them, European collectivism would rob British Thatcherites of the prize they had almost achieved after 13 years of struggle.

These fears, we postulated, would be combined with an earlier strand of Conservatism, highly evident in the backbench discontents of the early post-war years, which emphasised British independence and national sovereignty.[7] We considered it likely that, as in earlier periods, hostility to close involvement with Europe would be linked to a strong emphasis on law and order and severe penal policies.

For this study, we classified our demographic attributes into (i) *socialisation* variables, (ii) *constituency* variables and (iii) *generational* variables (see Appendix 1, p.68, for further details).

We first compared the make-up of the two broad groups – pro-Maastricht and anti-Maastricht – in terms of these attributes. What is striking is the way in which the composition of the two groups cuts across socialisation and constituency variables. Thus, there are no differences in the number of businessmen in the two categories, nor any differences in the sub-divisions of the business class. Company directors, whether public or private, were no more or no less likely to sign the anti-Maastricht EDMs. Indeed, occupationally the only noteworthy feature was the behaviour of the lawyers, who were almost three times as likely to sign anti-Maastricht motions as they were to be Europhiles.

Minor differences emerged among the different educational categories. Thus, the pro-Maastricht MPs were discernibly stronger amongst the alumni of the top 20 public schools (and especially so among those who went on to Oxford and Cambridge) than were their anti-Maastricht colleagues; the

Etonians for instance were twice as numerous amongst the pro-Maastricht group.

Constituency attributes show the same flatness as the occupational categories. None of the regions behaved distinctively. MPs from rural Britain reacted in the same way as urban members; pro-Maastricht members were somewhat more likely to come from constituencies where the Liberal Democrats were the main challengers.

One attribute alone stands out with clarity – parliamentary generation, namely the year of first election to parliament. One half of the pro-Europeans had been elected before 1979, compared with a mere quarter of the Eurosceptics (see Table 1). To put it another way, three-quarters of the Eurosceptics came into Parliament under the leadership of Mrs Thatcher or John Major. If anything, the Eurosceptics did better in the election of 1992 than in either 1983 or 1987; one third of them were returned in 1992, as against less than a fifth of our pro-Europeans.

Age reflects this pattern but in a blurred way. Older MPs were somewhat more likely to be pro-European, especially the over-60s, but the division is less clear-cut than for the parliamentary generations. It appears that date of election to parliament, rather than age, is the crucial factor although these two are of course correlated. The reason for this is that pro-Europeans coming from more privileged backgrounds would have tended to be younger, when first elected, than the anti-Maastricht MPs. Thus, if two MPs, one pro-European and one Eurosceptic, were elected for the first time in 1970, the pro-European would have been somewhat younger.

The importance of parliamentary intake reflects Conservative leadership attitudes to Europe. Most of the MPs who came into parliament before 1979 were of the Macmillan/Heath generations, who served their parliamentary apprenticeships and in many cases were chosen as parliamentary candidates when the Conservative Party was at the forefront of the drive for British entry into the European Economic Community (EEC). Europe became a badge of party orthodoxy, and the party's commitment was strengthened by the long battles in the country and later in parliament against a Labour Party,

TABLE 1
ATTITUDES TO EUROPE BY PARLIAMENTARY GENERATION

	First Entered Parliament					
	Before 1979		1979 or Later		Totals	
	No.	(%)	No.	(%)	No.	(%)
Pro-Maastricht	39	65	37	39	76	49
Anti-Maastricht	21	35	58	61	79	51
Totals	60	100	95	100	155	100

at best sceptical towards Europe and sometimes deeply hostile. Mr Heath's defeat as party leader, and his replacement by Mrs Thatcher, whose attitude to Europe was cooler and more distant than that of her predecessors, meant that reserve rather than enthusiasm became the hallmark of the party's stance.

Whiteley, Seyd and Richardson found a generational division of an ideological kind in their study of Conservative Party members, and offer a similar explanation to ours. '... Young Conservatives would tend to reflect Thatcherite beliefs, because these ideas were dominant in the party during their pre-adult years. By contrast, middle-aged Conservatives, whose pre-adult years were acquired in the Macmillan or early Heath period, are likely to be a lot more progressive, reflecting the dominant ideological tendencies within the party at that time.'[8]

How much this difference owed to the orientation of those coming forward for selection as Conservative candidates in different periods, to the selection committees themselves, or to encouragement from party headquarters is not clear.

The break between the pro-Europeans of the 1974 intake, and the scepticism of those first elected in 1979 is very sharp. There seems to have been a blip in 1983, when the number of pro-Europeans rose above the 1979 level (possibly connected with Labour's policy of outright withdrawal in 1983) followed by a decline in 1987. In the 1992 election, nearly two-thirds of the new intake were, by our measures, Eurosceptic;[9] but the behaviour of these members, the Major generation, warrants special consideration.

We have confirmatory evidence of this generational split from the signatories to a statement issued by the Positive European Group issued in January 1995. Fifty-two backbenchers signed this manifesto which called for Britain to 'press forward with confidence'. Of those who signed, 30 were first elected before 1979 and only five belonged to the 1992 cohort.[10]

One of the characteristics of a generational division is that it cuts across all other sources of cleavage. A strong preference for a specific policy by candidates at one election is likely to be reflected in all regions and so blur any underlying regional divisions. It will tend to have the same effect on socialisation indicators, such as education and occupation. Enthusiasm for Europe in the 1950s and 1960s cut across boundaries of school and university. An Etonian elected in 1959 or 1970 was likely to share the commitment of his non-Etonian colleagues, while one who entered the House in 1987 would tend to share the views of members who came into the House at the same time, irrespective of social background or occupational history.

The strength and clarity of the generational conflict, and the apparent lack of other explanatory cleavages prompted us to ask whether the split concealed significant divisions of other kinds. We examined, therefore, the relationship between demographic variables and opinion on Maastricht *within* each of our two major generations. In almost every instance we found that the dominance of the generational division survived intact. Thus, we find members from the most urban seats elected before 1979 opting heavily for Maastricht, whilst those who came into the House in or after 1979 are almost as heavily against. We find the same pattern in the rural and semi-rural seats.

The small size of some of the cells makes it hard to pursue this analysis into every demographic sub-group. As a generalisation, however, apart from some exceptions in the educational sub-groups, the generational cleavage is faithfully repeated.

Thatcherites and Others

The link between the wet–dry cleavage in the party, exposed by the policies of the Thatcher years, and the later division over Europe, has preoccupied the chroniclers of the Eurosceptic revolt. In their stimulating studies of the Maastricht rebellion, Baker, Gamble and Ludlam suggest that the key axes for understanding the Maastricht split (as well as the earlier divisions over the corn laws and tariff reform) are national sovereignty versus interdependence and extended government versus limited government.[11] The Wet/Dry split can easily be assimilated to the latter cleavage.

Baker and his colleagues note that the issue of Europe divided the Thatcherites as well as the wets; so Mrs Thatcher's first chancellor, Sir Geoffrey Howe, was separated both from the lady herself and from Norman Tebbit, a former Secretary of State for Industry.[12] However, these scholars tend to emphasise the free market/dry character of the Eurosceptic revolt. Thus, in their discussion of the 84 signatories of Fresh Start, they say that the third and largest group of these signatories were an assortment of 'free marketeers' or 'Thatcherites'. Their assessment of the 32 rebels and abstainers on the Paving Motion in November 1992 (presumably those who cared most about the issue) is that they 'were overwhelmingly characterised by their right-wing tendencies, and to a lesser extent their Thatcherite loyalties'. It is not altogether clear how they define 'right-wing tendencies'. Sowemimo, in what is in many ways an excellent appraisal of the division over Europe, argues that 'the sovereignty conflict has ultimately proved to be the decisive factor in Conservative ideological alignments'.[13] After repeating Baker *et al.*'s description of the majority of the Euro-rebels as being 'characterised by their Thatcherite and right-wing political loyalties', Sowemimo then claims that the 'Thatcherite character of this grouping (see footnote) is shown by the fact that the Baker and Spicer anti-federalist early day motions contain a large number of Tory MPs who are members of Thatcherite lobby groups such as the No Turning Back Group, the 92 Group and the Conservative Way Forward'. What we need to know here is how many, and what proportion.

We lay stress on, and discuss later, what may be called the authoritarian, or less emotively, the traditionalist elements in Conservative doctrine. The Euro-rebels consisted of those for whom sovereignty was an *instrumental* value and those for whom it was a *primary* value. For the former, the imperative to proceed with the Thatcher revolution requires the maintenance of British sovereignty just as, as Sowemimo observes, the anti-Market left in the Labour Party who opposed EEC entry in the 1970s saw 'the Community as a threat to their domestic economic ambitions'.[14] For anti-EEC socialists, the sovereignty of the nation state made possible, within British boundaries, the realisation of the socialist dream; for some of the Eurosceptic Conservatives of the 1990s, British sovereignty alone can ensure the success of the Thatcherite vision.

For some, but not for all, and perhaps not even for a majority.[15] For many Tories, sovereignty is an end in itself, to be kept at all costs, regardless of any temporary needs associated with particular political programmes. For such, the independence of the British nation is the mainspring of their creed. In a radio programme, Richard Shepherd conveyed the essence of this outlook.

> I think to many Conservatives – and I have to class myself in this group too – we're instinctive conservatives. It's the way we grew up, the way in which we look at the world, our communities and our relationships, etc., and the European question has challenged that … Conservatism is essentially about a sense of country in which each generation struggles to give that definition and coherence … so that belief in the institutions that we inherited, that we did not craft, that we are there to, in a sense, respect, conserve and change if necessary … is very central to this sense of country.[16]

Garry and Webb have carried out some highly interesting work using postal surveys to explore attitudes within the party to economic policy and moral issues and their relation to the Europe question. Their conclusions are informative but their utility for the present discussion is qualified by the time at which the research was undertaken. Opinion has moved on since the signature of the Maastricht Treaty. Indeed, we are conscious that it has moved on since the parliamentary debates of 1992–93. Nevertheless, their studies have carried forward the general argument about the roots of Euroscepticism in the Conservative Party. Garry, for example, used a postal survey to classify the policy beliefs of Conservative MPs in 1991.[17] He classified 70 per cent of his respondents as 'dries' and 30 per cent as 'moderates'. The dries were somewhat more Eurosceptic than Europhile, the moderates more heavily Europhile. On his figures, the cross-cutting tendencies of the split over Europe seem to be almost as significant as the Thatcherite credentials of the opponents of integration.

Paul Webb[18] carried out cluster analyses of MPs, on the basis of interviews undertaken by Norris and Lovenduski.[19] Their investigation covered the period April 1990 to October 1991, so Webb's findings are unlikely to be a precise guide to feeling in the 1992 Parliament. His figures showed that 69 per cent of the Conservative MPs surveyed favoured more integration with Europe, strikingly more than those Garry identified, who sent out his questionnaires late in 1991. Webb found majorities wanting more integration in all three clusters on the scale measuring attitudes to economic policy. However, the strongest support for closer links with Europe lay in the cluster he defined as the centre, a position corresponding roughly to the policies espoused by the Conservative wets.

It seems that opinion in the party has since moved sharply rightwards; Maastricht may have been the catalyst which mobilised latent anti-Europeanism in the party. However, Webb's work shows that, even before the crisis the Maastricht Treaty provoked, the wets were relatively more pro-European than their dry colleagues.

As we have seen, we concluded that our own data did not stretch to a reliable and meaningful wet–dry scale which would have enabled us accurately to classify backbenchers according to their opinions on economic and social matters. However, we had some EDM data which, we tentatively believe, allow us to throw some light on economic and social attitudes in the backbench Conservative Party.

In our dataset we included four EDMs calling for additional help for the disabled at home and two calling for action to help the Third World. Our 155 backbenchers between them provided 102 signatures; 57 signed at least one motion. To begin with, we separated out the two international-aid EDMs; these attracted a total of 32 signatures, all but three from our pro-Maastricht members. In short, Eurosceptics evinced little sympathy for the poor of the Third World. On the disability motions, 18 of the 79 Eurosceptics supplied 28 signatures amongst them; in contrast, 26 of the 76 pro-Europeans signed at least one of the disability EDMs, providing 42 signatures in all. This simple and limited measure of wet–dry sympathies lends some backing to the view that support for increased public welfare spending at home, and, *a fortiori* abroad, was greater amongst the pro-Europeans than amongst the Eurosceptics. Yet, there was some support among Eurosceptics for spending on domestic welfare measures, showing that opposition to the Maastricht Treaty could be found in the wet section of the party.

Traditionalism, Law and Order and Europe

Both Garry and Webb have examined the link between what may loosely be called moral questions and attitudes to European Union. Garry sought to establish the moral conservatism or moral liberalism of backbenchers with a single question on abortion. On this criterion, he reckoned that 51 per cent of the party were socially conservative and 37 per cent socially liberal.[20] As with his dry–moderate distinction, this cleavage was related clearly to opinions on Europe; his moral liberals, largely irrespective of their views on public expenditure, were strongly pro-European whilst his moral conservatives were preponderantly Eurosceptic. Among the latter, however, a negative attitude to public expenditure heavily reinforced a conservative moral stance.

Webb's cluster analysis, on the other hand, based on data from a period when support within the party for European integration may have been stronger, does not reveal any strong link between moral conservatism and attitudes towards Europe. The most interesting feature lies in the attitudes of what he calls the Libertarian right – those taking a permissive stand on moral questions who are hostile to higher public spending. Though still predominantly looking towards more integration with Europe, they were much less sympathetic than other moral libertarians, and indeed less so than his moral conservatives. It is among these libertarian dries that the true Thatcherite critics of European union are likely to be found.

Maastricht, the Death Penalty and School Punishments

The free vote on the death penalty, held in February 1994, and a vote on the reintroduction of caning in schools allow us to take further the argument about the link between moral conservatism and attitudes to Europe. Backing for severe legal punishments and to a lesser degree the physical punishment of children has often been linked in popular (and often hostile accounts) with a range of attitudes such as ethnocentrism, imperialism and intolerance – a compound which derives some justification from psychological research such as that by Adorno and his colleagues in the 1940s,[21] and from different standpoints by Eysenck[22] and more recently Glenn Wilson.[23]

Study of backbench Conservative attitudes in the 1950s and early 1960s showed that there was a connection between support for the death penalty and the restoration of judicial birching on the one hand, and a robust belief in empire on the other. Opinions on the Suez operation of 1956, and later on the Rhodesian question in the early 1960s, did tend to go with demands to keep capital punishment and bring back the birch. However, almost as striking were the vocal and conspicuous exceptions such as Angus Maude, Enoch Powell and John Biggs-Davison, who were all outspoken opponents of the death penalty.[24]

Two divisions took place on the death penalty in February 1994; we took the more general proposal – to restore the death penalty for all murders – which was the subject of a free vote, to see how attitudes to capital punishment varied with opposition to Maastricht. The pattern revealed in Table 2 is striking.

There is an emphatic contrast between the pro- and anti-Maastricht camps. A majority of the pro-Maastricht MPs voted against the restoration of the death penalty; the anti-Maastricht members, by a margin of three to one, voted to bring it back. Indeed, the high participation rate of the anti-Maastricht members is in itself notable. While there is no logical connection between attitudes to the European Union and capital punishment, there is a psychological link of some kind. Opinions about Europe seem to be related to feelings about the death penalty. This connection between pro-European and humanitarian attitudes can be identified in the late 1950s and was also present in the late 1960s.[26]

TABLE 2
VOTE ON DEATH PENALTY BY OPINIONS ON EUROPE

	Pro-Maastricht		Anti-Maastricht	
	No.	(%)	No.	(%)
For Death Penalty	27	46	51	76
Against Death Penalty	32	54	16	24
Totals	59	100	67	100
Not Voting	17		12	

Source: Parliamentary Debates. 23 February 1994

Thus far, our evidence is consistent with the findings of those psychologists who see support for severe policies on crime and punishment as being linked with ethnocentrism. We need to qualify this by noting the names of some prominent anti-Maastricht campaigners who declared against the death penalty – Nicholas Budgen, Michael Spicer and Bernard

Jenkin for example – and within the government such Eurosceptics as Peter Lilley, to whom we must add Jonathan Aitken, who sat in the cabinet until the re-election of John Major as party leader.

However, we also need to examine the demographic attributes of the two camps on the death penalty. Perhaps opposition to Maastricht, and support for the return of the death penalty are linked in other ways. Do opinions on capital punishment show a generational pattern? And how far do they reflect differing educational backgrounds? As a preliminary, however, let us explore the backgrounds of members in relation to the death penalty issue itself.

There is a rich harvest here. *Constituency* variables show that support for capital punishment was strongest among MPs from highly urbanised seats, from constituencies where Labour was runner-up and from north-west England, and weakest among MPs from the safest seats, and those in which the Liberal Democrats ran second. Scrutiny of the *socialisation* variables show two powerful educational effects. As in 1956, graduates were much more hostile to the death penalty than were their non-graduate colleagues. The difference was that in 1956 the abolitionist graduates were still only a small minority of Conservative backbench graduates.[27] In 1994, they mustered nearly half of all graduates voting on the issue. Moreover, in 1994 there was a striking link between the type of school a member went to and the way he voted on capital punishment. The more prestigious the school, the more likely he was to reject the death penalty, a finding reinforced by the behaviour of the small band of Etonians. Nor was this an artefact of the graduate/non-graduate division. It is true that all but a handful of the 57 members who went to one of the top 20 public schools went on to university and that only two-thirds of the backbench MPs from non-public schools did so, but nevertheless, it is clear that both school and university had independent effects. Non-graduates opted heavily for the death penalty, redbrick graduates less strongly and Oxford and Cambridge graduates went strongly against (see Table 3).

TABLE 3
VOTE ON DEATH PENALTY BY CATEGORIES OF SCHOOL AND UNIVERSITY

School	Oxford/Cambridge		Other University		No University		Totals
	For	Against	For	Against	For	Against	
Top 20	9	18	5	5	3	1	41
Other Public	12	18	20	13	13	4	80
Non-Public	10	5	23	8	14	4	64
Totals	31	41	48	26	30	9	185

Note: Percentages omitted from Table 3 for the sake of clarity.

Finally, the generational divide, which is so important in explaining the faultline over Maastricht, was also manifested in the death-penalty debate. Members first elected in the years up to 1974 broke almost equally in two; those entering the House from 1979 split almost two to one in favour of capital punishment. There was, in contrast, little difference among our four age groups, the impact of parliamentary generation being offset by the conservatism of age.

The generational cleavage is familiar to us; it may reflect a pronounced desire amongst Conservative activists over the last two decades to select candidates favourable to capital punishment. If so, it has been highly successful. To the extent that it has, it will have helped those other causes which supporters of capital punishment espouse. Eurosceptics may have been the incidental beneficiaries of severe views on law and order policy.

Attitudes to European integration, therefore, are linked to opinions on the punishment of crime. We have seen that views on the death penalty are sharply related to the type and length of education. What happens when we look at the two camps, pro-European and Eurosceptic, separately? Within each of these two groups, the influence of school and university was again highly marked; Oxford and Cambridge graduates, and the alumni of the top 20 public schools, again showed much less enthusiasm for the restoration of the death penalty than those who had not been to university and those who had not attended a public school. The absolute level of support for the death penalty was, of course, much higher among the Eurosceptics, but once again it varied sharply according to education.

Corporal Punishment in Schools

A second issue, which arguably elicits similar responses to the death penalty is the restoration of corporal punishment in schools. Caning in state schools was abolished in 1987, but in some parts of the Conservative Party there was a yearning for its return. Indiscipline in schools and bad behaviour among the young were regarded as the natural fruits of abolition.

In the last session of the 1992 Parliament, the government introduced an Education Bill. On Report stage, the Eurosceptic James Pawsey moved an amendment to allow the use of corporal punishment, under safeguards and with the consent of parents. The amendment was heavily defeated, Labour voted with the government and 99 Conservatives voted against the party whip.

Of those who had voted for the return of capital punishment, fully 83 per cent supported the reintroduction of caning in schools. Among opponents of the death penalty opinion was much more closely divided, with as many as 46 per cent in favour of corporal punishment in schools. Thus, though support for school caning and capital punishment do tend to go together among Tory backbenchers the proponents of these sanctions are very far from being a homogeneous bloc.

TABLE 4
VOTE ON CORPORAL PUNISHMENT IN SCHOOLS BY OPINIONS ON EUROPE

	Pro-Maastricht		Anti-Maastricht	
	No.	(%)	No.	(%)
For Restoration	16	29	40	62
Against Restoration	40	71	25	38
Totals	56	100	65	100
Not Voting or Not on Backbenches	20		14	

Source: Parliamentary Debates, 28 Jan. 1997

The division over corporal punishment in schools, like that over the death penalty, was clearly related to opinions on Europe with, for example, the former cabinet minister and challenger to John Major, John Redwood, voting with the restorationists.

The educational characteristics of those pressing for the re-introduction of corporal punishment in schools are similar to those of the MPs seeking the restoration of the death penalty. Fewer voted in the corporal punishment division than over the death penalty; moreover, opinion in the party was more favourable to the restoration of the death penalty than to corporal punishment in schools, and these differences may have blurred the clarity of the educational contrasts. Nevertheless, the Oxbridge graduates were distinguished from both non-graduates, and MPs from the other universities, by their heavy vote against school caning. Similarly, the public-school educated were hostile, while the MPs educated in state schools were strongly in favour, a reverse of the stereotypes to be found in the pages of such journals as the *New Statesman* in the 1940s and 1950s. As with the death penalty, it is clear that both school and university had independent effects.

One obvious question is whether the link between attitudes to the death penalty and views on European integration was, in a sense, spurious reflecting a generational difference. The pro-Maastricht members were drawn disproportionately from the longer serving (not necessarily the older) MPs; so, too, were the opponents of the death penalty. Is the link between the two issues an artefact of the year of entry into Parliament?

Our data speak loudly here. The Eurosceptics, with one important exception irrespective of their length of service in the House, supported capital punishment by a margin of three to one. It was their Euroscepticism as such, not their generational affiliation, which was related to their opinions on the death penalty. The pro-Maastricht members did tend to divide along generational lines, with the earlier cohorts declaring against capital punishment, and the Thatcher cohorts breaking narrowly in favour. However, the contrast between these later entrants and the Eurosceptics who came in in the same period is still stark.

Generational effects on the school caning amendment took a distinctive form with with the more recently elected being more heavily opposed than the pre-Thatcher intakes. However, within each generation pro-Maastricht MPs were much more hostile to the amendment than the Eurosceptics.

There remains one interesting quirk: in general, the Thatcher intakes were much more favourable to restoration of capital punishment than were those who entered parliament before 1979. However, the 1992 entrants (most of whom would have been selected as candidates when Mrs Thatcher was prime. minister) behaved distinctively. Among the Eurosceptics, only 10 per cent of those first elected between 1979 and 1987 voted against the death penalty; among the 1992 cohort, 43 per cent did. In the pro-Maastricht camp, just over a third of the 1979 to 1987 cohorts voted against restoration; among the 1992 intake, it was nearly two-thirds. Among the Eurosceptics, members first elected in 1992 provided 19 of the 25 votes against the amendment to bring back caning but only five of the 42 in favour. Our tentative conclusion is that the class of 1992 tended to be liberal on crime and punishment and related matters, but free-market and dry on economic questions and public expenditure. We suggest that the non-traditionalist Thatcherite element of Euroscepticism, for whom sovereignty is instrumental, not primary, can be found most clearly among this cohort.

THE EUROSCEPTICS AND THE FLOOR REVOLTS

So far we have sought to explore the links between ideology and rebellion over Maastricht. Another interpretation, canvassed at the time, focuses on much more mundane motivations – the career frustrations of MPs not yet promoted and the resentments among ex-ministers. Our data are particularly useful in throwing light on these alternative interpretations. EDMs are indicators of attitudes; votes in division lobbies are indicators of behaviour. Signing EDMs is generally a low constraint activity (despite exceptions such as Fresh Start, marked by higher constraint). Floor revolts, however, are subject to high constraint. Where the government makes a division one of confidence, constraint is near absolute.

We examined the participation of members in floor revolts in the light of their ranking on the anti-Maastricht scale. We hypothesised that some MPs would feel more constrained than others. Thus ex-ministers would consist of those who had little to lose and would therefore feel more free than other MPs to disobey the whip. Members who had been in the House since 1983 or earlier, and who had never held government office would also, we predicted, feel more relaxed about going into the lobbies against the government. These two categories comprise what John Major described as the 'dispossessed' (ex-ministers) and the 'never possessed' (the permanent backbenchers).[28]

Our third category consisted of the new parliamentary intake, the class of 1992. We call these the 'young hopefuls' – members recently elected and presumably anxious for government office. Such MPs, we argued, would be wary of open rebellion; the hope of appointment as a Parliamentary private Secretary (PPS) or of nomination to junior ministerial office would act as a powerful curb on any tendencies towards self-expression.

There were 47 past holders of government office in our dataset.[29] Of these, only nine were classified as anti-Maastricht. Given the importance of parliamentary generation this is not surprising. Most ex-ministers had first been elected before 1979, and would tend to be pro-Maastricht. Among the remaining 38 were a handful who rebelled once or more without having signed anti-Maastricht motions. John Biffen is the most conspicuous here. Of the 38, 18 were classified by us as pro-Maastricht and 20 were total non-signers. From this, we conclude that division-lobby rebellion is not a characteristic way for ex-ministers to express frustration. What stands out is that former ministers tended to hold themselves aloof from organised campaigning within the party. Of course, there were exceptions; we simply infer that the two largest groups among the 'dispossessed' were the pro-Europeans and the non-signers.

The 'dispossessed' and the 'never possessed', if we exclude the most recent intakes, comprise the whole backbench party. Thus we have no independent benchmark to set against the behaviour of these two categories. We can only compare them with each other, and with the 'young hopefuls'.

Fewer than half of the ex-ministers had a revolt score of more than five, and two never rebelled (see Appendix 1 for an explanation of revolt scores). The 36 permanent backbenchers, those who had never held office, reacted in a similar way to the ex-ministers. A quarter never rebelled at all, a half had a revolt score of more than five (see Table 5).

We chose the small intake of 1987 as occupying an intermediate position between the permanent backbenchers and the 1992 cohort. Some members

TABLE 5
REVOLT SCORES OF ANTI-MAASTRICHT MEMBERS OVER
SEVEN PARLIAMENTARY DIVISIONS[29]

	0	1–5	6–10	Over 10	Total
Ex-Ministers	2	3	–	4	9
Permanent Backbenchers	9	9	8	10	36
1987 Intake	2	1	2	3	8
1992 Intake	19	2	4	1	26
Totals	32	15	14	18	79

might feel frustrated at not having had recognition so far but still feel that time and good behaviour would bring their reward. In the event, the eight anti-Maastricht members of the 1987 intake proved to be a slightly more rebellious group than the permanent backbenchers.

The most striking behaviour came from those first elected in 1992, the 'young hopefuls'. We classified 26 of them as anti-Maastricht; 19 neither rebelled nor abstained on any of our key votes. Only one in five had a revolt score of more than five, and only one, Walter Sweeney, of more than ten. The behaviour of this group is a striking confirmation of our hypothesis. Members recently elected will be especially cautious about disobeying the whips. (E.g., three of the five signatories of the Fresh Start motion, EDM 174/92, who withdrew their names were from the 1992 intake.)

By the time of the Paving Motion debate in November 1992, some of the newly elected signatories of Fresh Start were having second thoughts. Thirteen backbenchers – 'members of the new intake of Conservative MPs' – wrote to the *Sunday Times* on 1 November to say that they would be supporting John Major in the Paving Motion debate in three days time. '... ratification of the Maastricht Treaty is essential for jobs and prosperity... A failure to ratify will remove Britain's ability to take a leading role in determining the future shape of the European Community.' Six of the signatories had signed Fresh Start in June.[31]

Three days later, on the day of the crucial debate, 15 of the 1992 intake who had signed Fresh Start wrote to *The Times*. Six had signed the letter published in the *Sunday Times* three days before. The new letter indicated that all 15 intended to vote with the government that night. Despite the overlap in signatures, the reasons given, however, differed considerably from those given in the *Sunday Times* letter. The Treaty as it stood made an 'encouraging start by redressing the drift to federalism'. Following the referendums in France and Denmark, and Britain's leaving the Exchange Rate Mechanism (ERM), events in Europe were moving to meet their own views. While they had reservations about certain terms of the Treaty, they did not think it right, at that time, to put at risk what had been achieved 'or to undermine the outstanding leadership of John Major'.[32]

There is, of course, an alternative explanation for the quiescence of the 1992 intake. What is surprising perhaps is not so much their conformity in the division lobbies, as their signing Fresh Start in the first place. Perhaps these MPs, who had been in the House for only six weeks when the motion was tabled, did not understand the significance of the motion they endorsed. Given the tactics, which by all accounts, were employed by the whips we incline to the view that most of the signatories did approve of the terms of Fresh Start. We are strengthened in this opinion by the failure of more than a handful of the 1992 intake to subscribe to the manifesto of the Positive European Group.

Our tables have shown that nearly half of those signing Eurosceptic motions were unwilling to express dissent in whipped divisions. In some cases, as we have seen, members' rationalisations, or reasons, were publicly voiced. Opinion on the European question bears, *inter alia*, the impress of political exigency and this is in parallel with the party's earlier experience.[33] In 1992 as in 1971, the lines between the two groups were not tightly drawn. On either side are dedicated partisans, but a considerable body of waverers and members unwilling to flout the whips can be found in between.

In accounting for the failure of many Eurosceptics to challenge the government in the division lobbies, we have emphasised constraints arising within Parliament. There are potential constraints outside the House coming from local Conservative associations. Here, like Baker, Gamble and Ludlam, we have been struck by the lack of visible pressures at this level. Some, such as John Wilkinson and Nicholas Budgen, have received warnings from their constituency associations, but most have had either the warm support, or at least the neutrality, of their activists. What we do not know though is how many of those who did not revolt were constrained by fears about the reaction of the local Conservative selectorate.[34]

One consideration which might influence a rebel member is the marginality of his constituency. Here, however, we can observe contradictory forces; an MP sitting for a marginal might well be concerned about his fate at the hands of the electorate if he and his fellow rebels were to bring about an election. On the other hand, as Leon Epstein has shown, MPs in marginal seats are less vulnerable to pressure from their constituency associations. The association in a marginal seat knows that its member may carry enough of a personal vote to hold the constituency in a tight contest; the MP sitting for a safe seat has no such sanction to employ.[35]

We can find a little evidence to support the thesis that fear of defeat at an early election inhibited some members from rebelling. Six out of the nine anti-Maastricht MPs sitting for the most marginal seats never rebelled at all. However, the 18 most persistent rebels were drawn more or less evenly from safest, middling and from the marginal seats. Walter Sweeney, with his lead of 19 votes in the Vale of Glamorgan, went resolutely through the dissidents' lobby with Teresa Gorman, basking in her 22,000 majority.

REFLECTIONS

Much has been made of the striking change that occurred in the composition of the parliamentary party during the Thatcher years – the displacement of the old social élites by new men and women from humbler backgrounds, educated at less prestigious universities and less exclusive schools.

We have found little evidence for these changes as such having had any major impact on the preferences of the backbench party on the main issues of government. On Maastricht, we found a limited tendency for the new classes to adopt Eurosceptic views. We emphasise, though, that these are differences of more and less. Indeed, there are signs that these social changes have had their greatest direct effects already on peripheral issues such as the death penalty and hunting. On capital punishment, for instance, changes in the educational background of the parliamentary party have had a discernible effect. Redbrick graduates have grown from 12 per cent of the backbench party to 38 per cent, and members educated at minor public and non-public schools have supplanted the Etonians and their allies from the top 20 public schools. On hunting, 30 Conservatives voted for the abolition of blood sports in the division on John McFall's bill. Only two of the 30 came from the top 20 public schools.[36] Reginald Bevins, Postmaster General in Macmillan's government in the early 1960s, commented that at one meeting at Chequers with Harold Macmillan and his colleagues, he and Ernest Marples were the only two without a landed estate – Bevins because he could not afford to, Marples because he had no use for one.[37] The 30 years that have passed span two different eras in the history of the Conservative Party; the social changes within the parliamentary Conservative Party have had little effect, however, on the central issues dividing the party.

However, natural turnover at the individual level has cumulated into major generational shifts, and these have had important indirect consequences. Many of the generation that carried through the momentous step of British adhesion to the EEC have either died or left Parliament. Whether as a result of the changing preferences of the pool of new, potential candidates or of constituency selection committees, they have been replaced by more Eurosceptic MPs.

Our second reflection relates to the place of the Maastricht faultline in the complex of the post-Thatcher, Conservative backbench party. Much comment has focused on the link between neo-liberalism, or Thatcherite economic policy, and hostility to Maastricht. We are cautious about classifying MPs as neo-liberals (or 'economic dries') on the basis of our present data, and while there is no doubt that some neo-liberals have been mobilised by the Maastricht issue, we think that the Maastricht revolt is better understood as a reflection of Tory traditionalism. The close connection between hostility to Maastricht and support for the death penalty is a persuasive witness to this judgement.

Our third reflection relates to the problems of party management which arise when a party has been in power for a long time. There is only a limited number of government posts to distribute, and the longer a government lasts, the more difficult it is to satisfy the clamour for office. The disappearance of the part-time member and the apparent increase in the numbers of those ambitious for promotion greatly exacerbate the problem.[38] Indeed, when a government has a large majority, as between 1983 and 1992, it becomes acute and the resentments which accumulated may help to account for Margaret Thatcher's defeat in 1990.[39] Conservative losses in 1992 helped to relieve the problem somewhat – but a fourth term, in a parliament likely to run for the normal four to five years, posed problems of its own.

A prime minister hoping to use the power of appointment to reward his followers finds himself in a dilemma. He can purge his government at intervals, by periodic reshuffles and appoint aspiring hopefuls in their place. In this way, he relieves the frustration of the recent recruits but at the risk of embittered senior colleagues accumulating on the backbenches. Moreover, as governments normally lose by-elections, he cannot avail himself of the option of rewarding the dismissed by sending them to the House of Lords, that humane Siberia of British politics, especially when his government has a narrow majority in the House of Commons.

Our figures indicate that the 'revolt of the dispossessed' has been exaggerated. Nevertheless, it deserves some credence; moreover, some very senior figures, not prominent in the Maastricht floor revolts, have since voiced their disquiet all too audibly: note Kenneth Baker's EDM of early 1995 attacking a single European currency and Norman Lamont's willingness to see Britain withdraw altogether from the European Union.

The resentments of the 'dispossessed', and the hunger of the 'never possessed' may have contributed to Margaret Thatcher's downfall in 1990. That event and the Maastricht rebellion offer a warning to governments which stay too long in office. There is a parallel here with the fall of governments in the Fourth French Republic and in post-war Italy. Changes of government may not betoken significant policy shifts, but they do provide an opportunity for a new share-out of the loaves and fishes of office. In the past, this function was performed in Britain by election defeats; moreover, under the British electoral system, a party's fall in support is reflected in a disproportionate loss of seats. Such bloodletting must help the leadership in its use of patronage.

Our last reflection is that this study has brought home to us the pitfalls of relying too heavily on floor revolts as a source of information about attitudes. Parliamentary divisions are usually carried out under high constraint. They confound two distinct elements – attitudes *per se*, and the propensity to behave independently, perhaps at personal risk. As we have seen, reliance on division-lists alone would have given a severely distorted picture of Euroscepticism in the Conservative Party. Free votes, of course, do not encounter this objection to the same extent.[40] In the last resort, all expressions of opinion are made under some constraint; the defect of the whipped division, as an indicator of underlying attitudes, is the severity of the constraint.

THE FUTURE

One of the biggest changes in recent British politics has been the Labour Party's transformation over the European issue. It is only 15 years since Labour called for British withdrawal from the EEC and was excoriated by the Conservatives for its irresponsibility. Today, it has to defend itself against the charge of embracing federalism. Meanwhile, the Eurosceptics speak of the need to repatriate power from Brussels to Westminster, and voices are heard calling for outright withdrawal from the European Union.

The sentiment of British nationhood has been the leading element in the belief-system of the Conservative Party during its evolution from the 1830s to the present day. Its manifestations have included imperial expansion and empire preference, hostility to the demands for greater autonomy from the peripheries of the United Kingdom, enthusiasm for the Ulster Protestants, support for the settler communities in East and Central Africa and, at an earlier time, the defence of the privileges of the national church. Nationhood, much more than market economics, has been the touchstone of the Conservative creed. The decline of British power and inexorably growing limitations on independent action have aroused acute fears. In so far as these beliefs merge with anxieties about the viability of the free market in the domestic economy, the result as Baker, Gamble and Ludlam[41] argue is likely to be explosive. However, we emphasise, more than they do, the independent force of the sentiment of nationhood.

The Conservative Party's troubles, of course, did not end with the ordeal of ratification. For the Major government, Europe remained a souce of deep conflict and continuing embarrassment. Slowly, the government was forced into successive concessions to the Eurosceptics, while sentiment in the parliamentary party gradually moved towards the sceptics. Meanwhile, the tide seemed to turn against the pro-Europeans within the constituencies. The arithmetic of retirement and replacement was against them. Only a quarter of the Eurosceptics had been elected before 1979, but a half of the pro-Maastricht MPs. It was not even necessary to postulate that retiring pro-Europeans would always be replaced by Eurosceptics to conclude that the party would soon have a clear Eurosceptic majority. The election of a Eurosceptic Leader in William Hague seemed to put the seal on the party's conversion to Euroscepticism.

The new Leader's first formulation of policy on European Monetary Union nevertheless made concessions to the erstwhile pro-European majority. The party proclaimed that Britain should not enter EMU 'for the foreseeable future' – a formula akin to that which enabled Mr Heseltine to stand for the leadership against Mrs Thatcher, without incurring the charge of inconsistency. The new Leader, however, like Pharaoh, then hardened his heart and ruled out entry for ten years, though at the cost of provoking a public protest from some of the party's most senior figures. Their demonstraation shows that the battle may not yet be over.

Given the continuing strength of traditional opinion among Conservatives, perhaps the true mystery is not why the Conservatives have, to use Mrs Thatcher's words about Labour in 1983, 'turned their back on Europe', but why they ever went into Europe in the first place. But that is another story.

APPENDIX 1

(A) THE DATA

(i) Data were collected for backbench members of the Conservative Party in the parliament elected in 1992, covering one full parliamentary session and the next up to 21 July 1994. All those holding government office during this period were excluded, though PPSs were included. Ex-ministers – defined as those who held government office before the general election of April 1992 but have been on the backbenches since the election – were included in the study.

(ii) Quantitative data in the study was analysed with Minitab, Version 8. These data were broadly of three kinds: demographic (individual background and constituency data), attitudinal (EDM signatures) and behavioural (the record of rebellion by backbenchers in the division lobbies).

(B) CONSTITUENCY VARIABLES

(i) Marginality

The winner's lead over the runner-up in the 1992 election was expressed as a share of the total vote, and coded in five steps from marginal to impregnable.

(ii) Second-Placed Party

The identity of the runner-up party was separately recorded.

(iii) Population Density

The average number.

(iv) Urban-Rural Character

The percentage of the workforce employed or self-employed in agriculture and related industries was noted for each constituency and then coded to produce a measure of the urban/rural character of the constituency, ranging from totally urban to very rural.

(v) Region

The region in which each MP's constituency is located was recorded using standard OPC definitions.

(C) SOCIALISATION VARIABLES[42]

(i) Gender

(ii) Education

Secondary and higher education were classified according to attendance at public school (top 20 and other), non-public school, and university (Oxford/Cambridge, other, none).

(iii) Occupation:

Most recent occupation before entering Parliament (in the case of officers in the armed forces, active service within the last five years). Where multiple occupations were given, company director trumped all others.

(D) GENERATIONAL VARIABLES

(i) Parliamentary Entry:

From year of first election to Parliament, MPs were grouped into parliamentary generations, to reflect cohorts elected at successive general elections (including intervening by-elections)

(ii) Age Groups

Members were coded into age groups, based on age in 1992 (e.g. MPs in their forties, MPs in their fifties).

(E) EARLY DAY MOTIONS[43]

(i) Any member of parliament may propose a Motion for Debate at an Early Day. An Early Day Motion (EDM), as such, has virtually no prospect of being debated. It is essentially an expression of opinion tabled by a member on any subject, to which any other MP may add his or her signature. A large number of EDMs were initially considered and discarded. Thirty-seven Conservative-inspired EDMs were selected and examined: this group was then whittled down to the 11 EDMs (shown below) chosen for detailed analysis.

Motions Analysed	Nature of Motion	Mnemonic
92/432	Welfare (Disabled)	19+Disab
92/174	Fresh Start re EU	FrshStrt
92/330	Welfare (Disabled)	Civdisab
92/549	Anti-ERM	FixdExch
92/621	International Humanitarian	HumAid
92/689	Economic Policy	EcPol

92/1253	Anti-EU	IntRatERM
92/950	Anti-EU	RusFed
93/2	Welfare (Disabled)	CRdisab
93/989	Welfare (Disabled)	IndMob
92/2099	International Humanitarian	G7AfAid

(ii) Principal Components Analysis (PCA)

The use of PCA was intended to disclose underlying statistical relationships within the body of EDM data. PCA is similar to factor analysis in that both methods enable the investigator to reduce a larger body of data to a smaller one. According to Kim and Mueller, PCA does not necessarily account for the observed correlations.[45] The quantitative results must be theorised and interpreted in the light of the researcher's substantive knowledge of the field.

(F) REVOLTS

The index of rebellion was compiled by scoring the number of occasions on which members defied the whips on seven divisions on the Maastricht legislation; MPs were scored one for deliberate abstention (where this was reported); two for voting against the whip. The divisions concerned were:

21 May 1992	Second Reading, European Communities (Amendment) Bill
	22 Tories oppose
4 Nov. 1992	Vote on the 'Paving Motion' to resume Committee Stage
	26 Tories oppose, 6 abstain
8 March 1993,	Committee Stage Vote on Amendment 28 (Council of the Regions)
	26 Tories oppose, 16 abstain (Government defeated)
22/23 April 1993	Committee Stage Vote on Referendum
	38 Conservative support, 8 abstain
20 May 1993	Vote on Third Reading
	41 Tories oppose, 5 abstain
22 July 1993	Vote on the Social Chapter 'to take note of the Government's policy on the Social Chapter'.
	23 Tories oppose, 1 abstention
22 July 1993	Vote on the Social Chapter (Labour Amendment)
	15 Tories oppose government, 8 abstain.

NOTES

Acknowledgements: We would like to thank Martin Harrop for his helpful comments on an earlier version. We also wish to thank Phil Appleby and Simon Brackenbury for their help and patience in recording the data and the University of Newcastle Upon Tyne Department of Politics for financial assistance. We have greatly profited from the illuminating, informative and often entertaining articles by Baker, Gamble and Ludlam. Paul Taggart and Paul Webb kindly furnished us with copies of past and forthcoming articles.

1. The title of Michael Spicer's book: *A Treaty Too Far* (London: Fourth Estate 1992).
2. D. Baker, A. Gamble and S. Ludlam. '1846 ... 1906 ... 1996? Conservative Splits and European Integration', *Political Quarterly* (1993) pp.420–34; D. Baker, A. Gamble and S. Ludlam, 'Whips or Scorpions? The Maastricht Vote and the Conservative Party', *Parliamentary Affairs* 46/2 (1993) pp.151–66; D. Baker, A. Gamble and S. Ludlam. 'The Parliamentary Siege of Maastricht 1993: Conservative Divisions and British Ratification', ibid. 47/1 (1994) pp.37–60.
3. S. Finer *et al.*, *Backbench Opinion in the House of Commons, 1955–59* (Oxford: Pergamon Press 1961); H. Berrington, *Backbench Opinion in the House of Commons, 1945–55*

(Oxford: Pergamon Press 1973); M. Franklin and M. Tappin, 'Early Day Motions as Measures of Backbench Opinion', *British Journal of Political Science* 7/1 (1977) pp.49–70; J. Leece and H. Berrington, 'Measurements of Backbench Attitudes by Guttman Scaling of Early Day Motions: a Pilot Study, 1968–69', *British Journal of Political Science* 7/4 (1977) pp.529–40; H. Berrington, J. Leece and N. Squirrell, *Backbench Attitudes in the British House of Commons 1959–76*, Report to SSRC, 1978.

4. s. fINER, IBID.
5. M. Burch and M. Moran, 'The Changing British Political Elite 1945–1983: MPs and Cabinet Ministers', *Parliamentary Affairs* 38/1 (1985) pp.1–15.
6. Strictly speaking, both types of indicator are attitudinal and both are behavioural. The critical distinction is between expressions of attitude given under low constraint, and those given under high constraint. Signature of an EDM is almost always cost-free, and rarely as highly-constrained as a vote cast against the whips. An adverse vote in the division lobbies is nearly always more than a token of opinion. Our distinction, then, between attitude and behaviour , if not made with perfect accuracy, reflects a substantive difference.
7. Finer (note 3).
8. P. Whiteley, P. Seyd and J. Richardson, *True Blues: The Politics of Conservative Party Membership* (Oxford: Clarendon Press 1994).
9. More strictly of those who signed one or more of our 11 EDMs.
10. This includes one signatory of Fresh Start, Geoffrey Clifton Brown, who signed no further anti-Maastricht EDMs.
11. Baker *et al.* (note 2).
12. Ibid.
13. M. Sowemimo, 'The Conservative Party and European Integration 1998–95', *Party Politics* 2/1 (Jan. 1996) pp.77–98.
14. Ibid.
15. P. Taggart. 'Rebels, Sceptics and Factions: Eurosceptics in the British Conservative Party and the Swedish Social Democratic Party', in I. Hampsher-Monk and J. Stamper *Contemporary Political Studies* Vol.1 (1996) pp589–97.
16. Richard Shepherd, in 'The Conservatives and Europe' (*Analysis*, BBC Radio Four, 6 April 1995).
17. J. Garry, 'The British Conservative Party: Divisions over European Policy', *West European Politics* 18/4 (Oct.1995) pp.170–89.
18. P. Webb, 'Attitudinal Clustering within British Parliamentary Elites: Patterns of Intra-Party and Cross-Party Alignment', *West European Politics* 20/4 (Oct. 1997) pp.89–110.
19. P. Norris and J. Lovenduski, British Candidate Survey. Reported by the authors, in *Political Recruitment: Gender, Race and Class in the British Parliament* (Cambridge: CUP 1995).
20. Garry (note 17).
21. T. Adorno *et al.*, *The Authoritarian Personality* (NY: Harper 1950).
22. H. Eysenck, *Psychology of Politics* (London: Routledge 1954).
23. G. Wilson (ed.), *Psychology of Conservatism* (London: Academic Press 1973).
24. Finer (note 3).
25. Ibid.
26. Berrington (note 3).
27. Finer (note 3). Note that Whiteley, Seyd and Richardson found that support for capital punishment and opposition to further European integration had loadings on the same factor in their factor analysis of Conservative Party members. Whiteley *et al.* (note 7).
28. *Sunday Times*, 25 July 1993.
29. Excluding John Horam, who was a junior minister in the Labour government of the 1970s and was appointed to junior office in John Major's government.
30. The 79 include only those defined as anti-Maastricht on our EDM scale, and the revolts only those mentioned in the Appendix. In addition to the 79, John Biffen (total non-signer) rebelled on the Maastricht bill (score of nine). George Walden, John Stanley and David Howell were reported as abstentions in the division of 8 March on the Committee of the

Regions – they like Biffen being total non-signers. The abstentions of Walden and Howell, given their general stance, were specific to this issue. Kenneth Baker, a non-signing ex-minister, rebelled during the Maastricht bill's progress though not on any of the divisions we have used (see table by D. McKie and R. Leonard, *Guardian*, 5 May 1993).

31. *Sunday Times*, 1 Nov. 1992.
32. *The Times*, 4 Nov. 1992.
33. See, for instance, Kitzinger's analysis of Conservative behaviour over the vote in 1971. U. Kitzinger, *Diplomacy and Persuasion: How Britain Joined the Common Market* (London: Thames & Hudson 1973).
34. See *The Times*, 10 March 1993.
35. L. Epstein, *British Politics and the Suez Crisis* (London: Pall Mall 1964).
36. Parliamentary Debates, March 1995 Vol.255 Col. No.1367–68.
37. R. Bevins, *The Greasy Pole* (London: Hodder 1965).
38. A. King, 'The Rise of the Career Politician in Britain – and its Consequences', *British Journal of Political Science* 11 (July 1981) pp.249–86.
39. This view was expressed by Sir Bernard Ingham at a private seminar.
40. MPs may be subjected to much more intense lobbying from pressure groups on free votes, of course.
41. Baker *et al.* (note 2, 1994 p.53) In this article, they refer to the 'inflammable combination of these tensions'.
42. Based on information supplied in *Times Guide to the House of Commons* and other reference sources. Our classification of schools followed that by R. Kelsall, *The Higher Civil Service in Britain* (London: RKP 1955).
43. *Parliamentary Papers*, Notices of Motions.
44. J. Kim and C. Mueller, *Introduction to Factor Analysis: What it is and How to do it*, Sage University Paper, No.13 (London and Beverly Hills: Sage 1978); J. Kim and C. Mueller, *Factor Analysis: Statistical Methods and Practical Issues* Sage University Paper, No.14 (London and Beverly Hills: Sage 1978).

From Hostility to 'Constructive Engagement': The Europeanisation of the Labour Party

PHILIP DANIELS

THE LABOUR PARTY'S CHANGING STANCE ON EUROPEAN INTEGRATION

Since the early 1960s, Europe has been a source of division and difficulty for the Labour Party. The party's stance on European integration has oscillated between outright opposition and positive support, with each shift in policy shaped by intra-party factional conflicts, domestic political competition and the dynamics of the integration process. Even during periods in the 1960s and 1970s when official party policy favoured Britain's membership of the European Community (EC), the support was not wholehearted and the issue tended to divide the party at all levels.

When the first steps were taken towards an integrated Europe in the early 1950s, the Labour Party was hostile to Britain's participation in the process. At the time of Britain's first application for membership of the Common Market in 1961, Labour, then in opposition, abstained in the Commons vote on the principle of Britain's entry. The Labour government elected to office in 1964 showed little immediate interest in the issue of Britain's membership. By the 1966 election, however, the party included a manifesto commitment to seek entry into the European Economic Community (EEC) provided essential British and Commonwealth interests were safeguarded. The government's application for British entry was vetoed, however, by President de Gaulle.

Labour's conversion to a pro-membership stance provoked much conflict within the party and the issue continued to disrupt the party in government and in opposition until the late 1980s. In the 1970 election, Labour remained committed in principle to Britain's entry into the EC. The election was won, however, by the Conservatives and the new government, led by Edward Heath, applied for British entry in 1971. The divisions in the opposition Labour Party over the European issue quickly surfaced. In October 1971 Labour voted in the House of Commons against entry into the EC, but 69 Labour MPs defied a three-line whip and supported the

Conservative government's position on British accession. Faced with a serious problem of party management, Harold Wilson, the Labour leader, committed the party to renegotiate the terms of entry agreed by the Conservative government and to put the new terms to the British people in a national referendum. Following the Labour Party's return to office in 1974, the new Labour government renegotiated the terms of entry, and the promised referendum, held in June 1975, saw a large majority in favour of Britain's continued membership of the EC. The Labour Party remained deeply split, however: while a majority of the cabinet endorsed the new terms, they were rejected by a small majority in the Parliamentary Labour Party (PLP), by the National Executive Committee, by the constituency Labour parties and by the trade unions.

The referendum result did not finally resolve the issue of Britain's membership and divisions persisted in the Labour Party. Labour's period in office from 1974–79, under the premierships of Wilson and James Callaghan, was characterised by a lack of enthusiasm for European integration. Following the party's defeat in the 1979 election and the election of the anti-EC Michael Foot as leader in 1980, Labour's attitude towards Europe became increasingly hostile. The 1980 Labour Party conference passed a resolution advocating Britain's withdrawal and at the same time ruled out future negotiations since the party believed that the Community was incapable of changing in a way that would make continued membership acceptable for Labour.

The party's adoption of a strongly anti-EC policy encouraged leading pro-European moderates to break away from the party and form the Social Democrat Party (SDP) in 1981. In the 1983 election, the manifesto committed the party to a policy of withdrawal from the EC, a pledge to be implemented within the lifetime of a parliament.

Throughout this period from the late 1950s to the mid-1980s, of the two major parties, the Conservatives were consistently more pro-European than the Labour Party. Labour was deeply divided over the European issue and even during periods when official party policy was pro-EC, its support was based primarily on a pragmatic economic assessment of the advantages to Britain rather than on a deep commitment to the process of European political and economic integration.

Labour's difficulties with European integration derived from its commitment to Britain's existing global links, its analysis of the EEC and its focus on the nation-state as the appropriate arena in which to achieve its policy objectives. In the 1950s, the Labour Party shared the Conservatives' concerns that participation in the integration process would undermine Britain's defence and security links with the United States and its important trade relationship with the Commonwealth. The Labour Party also had misgivings about the nature of the EEC which it regarded as a capitalist club for the rich and an insular organisation. Most importantly, Labour feared that British membership of the EEC, whose member-state governments were dominated by parties of the right in the 1950s and 1960s, would frustrate attempts to carry through socialist policies in Britain. Concerns about the loss of national sovereignty inherent in the process of European integration have been a recurrent theme in Labour difficulties over the European issue. For the anti-European left of the party, the retention of national economic sovereignty was the principal factor in their opposition to British membership. On the right of the party too, a small anti-European element opposed membership largely on the grounds that it would undermine parliamentary sovereignty.

Attitudes towards Europe within the Labour Party became entangled with intra-party factional conflicts over the control and ideological direction of the party. Labour's European policy was often ambiguous and contradictory and the vacillations between pro- and anti-positions reflected in part the interplay and changing balance of left and right forces in the party. With deep divisions over Europe at all levels of the party, the leadership faced persistent and often intractable problems of party management. The shifts in Labour's European policy were also caused by the adversarial nature of political competition in Britain. The European question has rarely been an issue of bipartisan consensus in post-war British politics. For example, Labour's strong opposition to EEC entry on 'Tory terms' in 1973 was motivated in part by its deep hostility towards the economic and industrial relations policies of the Heath government.

THE EVOLUTION OF LABOUR'S PRO-EUROPEAN POLICY, 1983-97

Between 1983 and 1987, Labour slowly shifted away from outright rejection of Britain's EC membership towards an unenthusiastic acknowledgement that early withdrawal was not a viable policy option. Labour's defeat in the 1983 election was the worst in the party's history and led to a review of a whole range of policies. The party's commitment to withdraw from the EC had appeared negative and unrealistic, providing the Conservatives with an easy target during the election campaign. With the election of Neil Kinnock as party leader in 1983 a new phase began in Labour's stance on the European issue.

The initial softening of the party's anti-EC position was a pragmatic response required by two factors: first, the need to prepare for the 1984 European parliamentary elections, in which the Labour Party hoped to

TABLE 1

SUMMARY OF LABOUR POSITIONS ON THE EUROPEAN ISSUE, 1975–1997

1975	Labour government renegotiates limited changes to Britain's terms of membership. The cabinet supports Prime Minister Wilson by 16 votes to 7 and endorses continued British membership of the EC under the revised terms. However, 18 out of 29 NEC members reject the new terms and a special party conference calls for a 'no' vote in the national referendum on the issue. In the Commons debate in April 1975, 38 ministers vote against the government, 9 abstain and 45 vote in favour. Among backbench Labour MPs, 107 vote against the leadership, 24 abstain and 92 vote in favour.
1976	NEC document opposing direct elections to the European Parliament (EP) is passed by the Labour Party annual conference but without a sufficient majority to make it party policy.
1978	Labour government declines to take pound into the ERM of the European Monetary System.
1980	The party's annual conference passes a resolution advocating withdrawal from the EC.
1981	The annual conference reaffirms commitment to withdraw with no further referendum on the issue.
1983	Labour's election manifesto includes the commitment to withdraw from the EC 'to be completed well within the lifetime of the parliament'.
1984	Labour's manifesto for the elections to the EP calls for fundamental reform of the EEC.
1987	Labour's election manifesto makes no mention of the withdrawal option.
1988	Launch of the Policy Review exercise which formalises Labour's pro-European stance.
1989	Labour presents itself as the pro-European party in the election for the EP.
1990	NEC statement indicates that Labour would support a single currency and a politically-accountable European central bank.
1993	Labour supports the Maastricht Treaty but criticises the Conservative government for its opt-out from the Social Chapter.
1997	Labour's policy of 'constructive engagement' with the EU is fully reflected in the party's election manifesto.

reassert itself as a credible alternative to the Conservative Party; and second, since Labour's election defeat in 1983 had ruled out withdrawal from the EC in the short term, the party needed to adapt to that reality and shift its focus to the issue of reform of the institutions and policies of the EC.

The Labour Party's initial reorientation on the European issue in the post-1983 period lacked clarity and consistency. While there was a subtle shift away from the manifesto commitment to withdrawal from the EC, there was little positive enthusiasm for the integration process and the party retained the option to withdraw if the EC failed to undertake fundamental reform of its policies and institutions. The party's manifesto for the 1984 European parliamentary elections ('A Fair Deal for Britain and a New Deal for Europe') reflected Labour's equivocal stance on the European question. The manifesto played down the party's traditional hostility to the EC and acknowledged that Britain's immediate future was bound up with Europe. There was, however, little positive endorsement of the EC: on institutional issues, the manifesto stated the party's opposition to any increase in the legislative role of the European Parliament and demanded the return of powers to the Westminster Parliament; on policy issues, the manifesto called for fundamental reforms of the Common Agricultural Policy (CAP) and a reduction in Britain's budgetary contributions.

The 1987 Labour Party manifesto devoted a single paragraph to European policy:[1] it made no mention of the withdrawal option and continued the shift away from outright opposition to the EC towards a policy which focused on the need to reform the EC and to use it as a vehicle to advance Britain's interests. In effect, this represented an implicit, if reluctant, acceptance by the party that continued membership was the only viable option for Britain.

The Labour Party's more overtly positive approach to European integration developed as a part of the policy review undertaken by the party following its third successive general election defeat in 1987 with a 30.8 per cent share of the vote.

> ... the smooth professionalism of Labour's campaign had won acclaim – but the Party was not spared another crushing defeat at the polls. Not only were there several points at which it was still at odds with popular feeling, but the pervasive lack of trust in the Party, the lack of confidence in its ability to govern and manage the economy competently all convinced the leadership that a major overhaul of policy was essential.[2]

The Labour leader, Neil Kinnock, pressed for a fundamental reappraisal of the party's doctrine and programme in an attempt to widen its electoral appeal and to challenge the Conservative Party's dominance of national government. The Policy Review exercise, launched in 1988, consisted of seven policy review groups, made up mainly of representatives from the shadow cabinet and the NEC.[3] Four Policy Review reports were published between 1988 and 1991 and innovations in Labour policy on a whole range of issues were introduced incrementally over this period.[4]

Each of the reporting groups in the policy review exercise addressed the European dimension, a clear indication of the growing importance which the party leadership attached to the European issue and an acknowledgement that Europe provided the essential framework for co-operation over a range of policy concerns. The Policy Review accepted the Single European Market and called for co-ordinated policies among members states on environmental protection, workers' rights, social benefits and reflationary programmes. The review supported European Political Co-operation in foreign policy, but rejected any notion of Europe developing as a superpower. It recognised the lack of democracy and accountability in the EC's decision-making processes and, without giving specific detail, called for decisions to be taken at the level where democratic control and effectiveness would be maximised. The review supported the retention of the national veto in EC decision-making and called for enhanced powers of scrutiny for the British Parliament in dealing with EC legislation. Thus, in terms of European policy, the message of the policy review was that the EC was an appropriate arena in which to advance British interests. The second policy-review report, *Meet the Challenge, Make the Change*, which contained the core of the pro-European policy, was approved by the 1989 annual conference.[5]

The more positive stance on Europe embodied in the policy review served as a symbolic break with the policy of withdrawal contained in the 1983 manifesto and it served to highlight a weakening of the 'hard left' in the party's internal balance of power. At the same time, Labour's official pro-European stance enabled it to criticise the Conservative government's policy on Europe and to fill a political space, based on constructive Europeanism, once occupied by the Conservative Party but increasingly vacated by it after 1988.

In the 1989 elections to the European Parliament, the Labour Party abandoned its anti-EC stance in favour of a more positive approach to Europe. It called for a strengthening of EC institutions, greater democratic accountability and increased co-operation in the field of social policy. With the Conservative Party taking a much more sceptical view of the EC's political and economic direction, and fighting an essentially negative campaign in the Euro-elections, Labour was able to present itself as the more pro-European of the two major parties. Labour emphasised the potential benefits of EC membership for Britain and criticised the Conservative Party's anti-EC stance which, it claimed, left Britain increasingly isolated in Europe and undermined its national interests.

Labour's pro-European orientation did not weaken in the 1990s. A commitment to the pound's entry into the ERM became official party policy in 1989 and the party supported Britain's ratification of the Maastricht Treaty in 1993.[6] The election of John Smith as party leader, following Labour's fourth successive election defeat in 1992, strengthened the party's pro-European stance. Smith, a consistent pro-European throughout his parliamentary career, secured an overwhelming victory against Bryan Gould who was a strong critic of key aspects of Labour's European policy and in particular its commitment to sterling's membership of the Exchange Rate Mechanism (ERM). In the 1994 elections to the European Parliament, the party's manifesto ('Making Europe Work for You') focused on the need for Britain to take advantage of the opportunities offered by EU membership. The Euro-election campaign was overshadowed, however, by John Smith's death shortly before polling. His successor, Tony Blair, has continued Labour's pro-European stance in opposition and in government since May 1997. He has emphasised the party's commitment to 'constructive engagement' with the European Union (EU) as the best way to advance Britain's interests.[7] At the same time, the party under Blair has moved cautiously on European issues such as the single currency and institutional reform in order to thwart Conservative attempts to depict Labour as a 'soft touch' or 'Brussels' poodle'.

LABOUR'S CONVERSION TO EUROPEANISM

During the 1980s Labour's approach to Europe moved from deep hostility, through studied inattention to positive acceptance. Labour's shift to a pro-European stance has no clear point of origin. It evolved steadily in the post-1983 period, initially as a political necessity but later embraced by the party as an opportunity to advance its political and economic goals.

Labour's repositioning on the European issue was carried through incrementally under the leadership of Neil Kinnock. Between 1983 and 1987, the party slowly extricated itself from the withdrawal commitment and focused on the need to reform the EC from within. After the 1987 election defeat, the development of a pro-European policy became a central part of the Policy Review exercise. The incremental way in which the party's European policy was remodelled reflected the leadership's initial difficulties in countering the influence of a united and organised left in the party. From the mid-1980s, however, the influence of the left weakened as 'polarization began to unravel, cleavage patterns became more complex, lines of demarcation increasingly cut across each other, and attachment to factional groupings abated'.[8] With the support of the party's 'soft left' and the trade union leadership, Kinnock was able to construct a coalition for modernisation and get a tight grip on the party's key centres of power.

Control over policy was shifted towards the PLP and in particular the shadow cabinet. Kinnock describes the shift in Labour's European policy as one which could 'be changed without great resistance'.[9]

> Sometimes I could rely on advocacy by myself and other Shadow Cabinet and Frontbench colleagues mixed with the effect of the passage of time and events which patently made established policy obsolete. The movement of policy on the European Community, which I had started before becoming leader, is a case in point.[10]

Simultaneous and related developments in domestic politics and in the European arena explain the transformation of Labour's position on Europe. The pressures of domestic political competition, a change in trade union attitudes on Europe, the dynamics of the European integration process, and important changes in the party's approach to economic policy and the role of the nation-state are key elements in understanding Labour's new-found Europeanism.

DOMESTIC POLITICAL COMPETITION AND ELECTORAL STRATEGY

The Labour Party's gradual shift to a pro-European stance in the post-1983 period owes much to domestic political considerations. The party's repositioning on the European issue has passed through various phases and, at each, the domestic political context and internal party politics have been important in shaping Labour's strategy. In general terms, the 'Europeanisation' of the party may be seen as a response to its long exclusion from national office and as a key element of a broader electoral strategy designed to convey the image of a party which is modern, credible and fit to govern. The policy review exercise, launched in 1988, was central to this strategy of programmatic renewal.

Crushing defeats in both the 1983 and 1987 elections convinced the party leadership that a major renewal of policy was essential. Labour's incremental move to a pro-European stance became a fundamental element in this overhaul of party policy. The Labour leadership regarded the 1983 withdrawal commitment as damaging to the party's electoral prospects since it contributed to an image of a party which was out of touch, lacking credibility and dominated by the left. In its moves towards a pro-European position, the party leadership sought to improve its overall image as a party fit to govern and to reflect majority British opinion which favoured continued membership of the EC, but generally showed little enthusiasm for deeper political and economic integration.[11]

The pro-European stance thus became a key element in the Labour

Party's move back towards the middle ground of British politics. The gradual embrace of Europeanism helped the party to recapture some of the political ground lost following the formation of the SDP in 1981; it has symbolised the party's shift away from the left-wing programme which characterised its early years in opposition in the post-1979 period; it has enabled the party to highlight and exploit the deep divisions in the Conservative Party over Europe and to claim that the Thatcher and Major governments' self-inflicted isolation in Europe has been damaging to British interests; and its support for the pound's entry into the ERM and, in principle, the single currency has reinforced the credibility of the party's commitment to fiscal discipline and responsible economic management.

In the initial stages of Labour's evolution towards a more positive Europeanism under Kinnock's leadership, the party was concerned to recapture some of the political ground lost to the strongly pro-European, Liberal-SDP Alliance. By the late 1980s, however, the European issue was becoming increasingly salient in the political competition between the Labour and Conservative parties. The Conservative Party's increasing unease with the pace and direction of European integration from the late 1980s onwards gave the Labour Party an opportunity to make political capital from the European issue. Mrs Thatcher's Bruges speech of September 1988, with its explicit Eurosceptical message, marked a turning point for both the Conservatives and the Labour Party on the question of Europe. For the Conservative Party, Thatcher's growing hostility to the European project exposed deep divisions in the party which were to play a significant part in her replacement as prime minister in November 1990.

Her successor, John Major, attempted initially to pursue a more positive European policy designed to put Britain 'at the heart of Europe'. Major found himself hemmed in, however, by the Eurosceptical elements of the Conservative parliamentary party. Major's problems of party management over the European issue became quickly apparent during the Maastricht Treaty negotiations of 1991 and were to become more acute with the Conservatives' reduced Commons' majority after the 1992 election and sterling's ignominious exit from the ERM in September 1992. The Conservative Party's deep and visible fissures over the issue of European integration presented Major with severe problems of party management throughout his period in office. His attempts to appease both sides of the European divide in the party merely exacerbated the clear lack of direction on key policy areas such as the single currency.

The Conservatives' increasing difficulties over key aspects of Europe's political and economic integration helped to reinforce Labour's own pro-European stance. On the issue of the 'social dimension', in particular, which became a key area of dispute between the Conservatives and Europe, the Labour Party and the Trades Union Congress (TUC) aligned with the European position. Labour's claim to be the party of Europe, an important element in its 1989 Euro-election campaign, appeared justified. The rise of Euro-scepticism in the Conservative Party, and the party's efforts to put 'clear blue water' between itself and Labour, opened up a political space on the European issue which the Labour Party was ready and willing to occupy. There was, in reality, much common ground between the two major parties on many of the key European issues but this was largely concealed as a result of Major's attempts to placate the Eurosceptics in his party. From the late 1980s onwards, the Labour Party was able to capitalise on the Conservative Party's divisions over Europe in at least three ways. The lack of consistency in Conservative European policy gave the Labour Party an opportunity to attack Major for weak leadership; Labour was able to portray the Conservatives as ideologically and politically isolated in Europe in contrast to Major's declared aim to put Britain at 'the heart of Europe'; and Labour was able to claim that Major's attempts to pacify the Eurosceptical wing of his party was undermining Britain's position in Europe. For the Conservatives, just as with the Labour Party in the early 1980s, the European issue exposed deep internal divisions and damaged the party's image.

By contrast, since the late 1980s Labour has projected an image of a party largely united over Europe and with little significant ideological opposition to Britain's European membership. Labour's approach to European policy during this period has been characterised by a cautious pragmatism. The party has emphasised that 'constructive engagement' with the EU rather than a negative, confrontational approach is the best way to advance British national interests. It has supported key European policy initiatives, such as the 'social dimension'[12] and Economic and Monetary Union (EMU), which have been politically difficult for the Conservative Party. At the same time, however, Labour's positive Europeanism has been balanced by an undertaking to defend British national interests (described by Blair as 'the patriotic case for Europe'[13]) and to oppose the creation of a federal EU, thus making it difficult for the Conservatives to gain a significant electoral advantage over Labour on the European issue.[14] This caution was clearly evident in the run-up to and during the 1997 election campaign: while Labour maintained its essentially pro-European stance, on issues such as the single currency and treaty reform it ensured that there was little ground between it and the Conservative Party.[15]

The controversy over Britain's membership of an EMU illustrates clearly the different ways in which the Conservative and Labour parties have handled the European issue in the 1990s. For the Conservatives, the 'wait and see' formula on the single currency issue produced an uneasy truce within the party which it was unable to sustain throughout the long 1997 election campaign. The widespread opposition in the Conservative Party to Britain's membership of a single currency is based principally on political and constitutional concerns relating to fears of a loss of national sovereignty. The Labour Party, in contrast, has sought to treat the single currency issue as an essentially economic question.[16] In the run-up to the 1997 election, Labour's policy on the single currency became virtually indistinguishable from that of the Conservative government. It would 'wait and see' if and how the single currency would be set up. It acknowledged that 'formidable obstacles' made it unlikely that Britain would participate in the first wave of members of the Euro zone and it promised a referendum if a Labour cabinet recommended and Parliament agreed to the pound's participation in a future single currency (the so-called triple lock).

   In spite of this common ground between the two major parties, an important and revealing difference remained: given the extent and depth of the ideological opposition in the Conservative Party, it appeared unlikely that a future Conservative government would be able to agree to the pound's entry into a single currency, even if an economic case could be made for Britain's participation. The Labour Party's position appeared more flexible and its policy options more open; a future Labour government would base any decision on the pound's participation on a pragmatic, 'hard-headed economic assessment' of Britain's national interests. While elements of the PLP remain against the single currency, their opposition is typically based on what they regard as the inherent economic flaws in the EMU project and the unacceptable constraints it would impose on the Labour government's room for economic manoeuvre rather than the constitutional concerns[17] which are at the heart of Conservative opposition to the single currency. The divergence in the two parties' positions on the EMU issue was confirmed in October 1997: the Conservative Party ruled out British membership of the single currency for at least two parliamentary terms while the Labour government committed itself to the principle of the pound's entry but would join only if an economic case case could be made for membership.

ECONOMIC MANAGEMENT IN THE EUROPEAN CONTEXT

Changes in Labour's economic policy since the early 1980s have been an important factor in the party's accommodation with Europe. There are several elements in this link between the party's Europeanisation and changes in its economic thinking. First, the acknowledgement across the party that growing economic interdependence imposed significant constraints on national economic strategies made it easier for Labour to come to terms with membership of the EC. Second, a commitment to Europe was a key part of Labour's efforts to enhance its credibility as a party of competent economic management. Third, since the late 1980s there has been a growing congruence between Labour's macro-economic policy and the economic doctrine at the heart of the EU's moves towards EMU.

Economic policy was a central factor in the Labour Party's anti-European position in the early 1980s. The Alternative Economic Strategy (AES), to which the party was committed in the early 1980s, was based on a mixture of protectionism, assistance to national industries, renationalisation and expansionary Keynesian spending programmes. Much of this proposed programme was incompatible with the obligations of EC membership. The 1983 manifesto commitment to withdrawal from the EC was, in part, an acknowledgement that membership would place severe constraints on the implementation of a future Labour government's programme based on the AES.[18] Given the British economy's deep enmeshment in Europe, however, withdrawal from the EC appeared unrealistic and damaging to Britain's long-term economic interests.

The evolution of Labour's pro-European stance has gone hand in hand with the abandonment of the nationalist approach at the heart of the AES and the reshaping of the party's economic policy to one based on acceptance of the market economy, orthodox monetary and fiscal policy, and Britain's full participation in an open, global economy. The Europeanisation of the Labour Party has been eased by the party's acknowledgement of growing economic interdependence and the futility of an economic strategy based on the concept of insulating the domestic economy from the constraints imposed by the international economic context within which Britain operates.

The experience of the Mitterrand government in France between 1981 and 1983 demonstrated the international economic constraints operating on national governments. The French Socialists' attempts to reflate the economy, at a time when its European partners were pursuing deflationary policies, ended in failure and an enforced policy 'U'-turn. The 'Mitterrand experiment' showed clearly that reflation in one country, as Labour's AES proposed, was no longer a feasible policy option. The growing interdependence of European economies, the mobility of capital and liberalised trading arrangements rendered obsolete purely–national Keynesian reflationary strategies. For the Labour Party, the failure of the French Socialists' reflationary strategy highlighted the inherent flaws in its own AES. The pro-European elements of the Labour Party had long recognised the limits of domestic economic strategies and the advantages of co-ordinated EC-wide economic policies. By the mid-1980s sections of the Labour left, traditionally the faction in the party most hostile to European integration, were moving away from an economic policy centred on nationally based reflationary strategies towards an approach that focused on the EC as the appropriate arena in which to pursue Labour's economic objectives.

The Labour Party's move to a pro-European stance may be viewed as part of a strategy to reassure the electorate and key business and financial interests that the party is committed to responsible and competent economic management. There are two elements to this: first, in marked contrast to the 1983 withdrawal commitment, the party's pro-Europeanism represents an acknowledgement that Britain's economic future is inextricably bound up with the EU; second, since the late 1980s the party has been willing to accept the external disciplines and constraints on monetary and fiscal policies imposed by the EC/EU in the form of the ERM and the convergence criteria for the single currency.

The issue of the pound's entry into the ERM divided Labour in the late 1980s. The report for the policy review process, *Meet the Challenge, Make the Change* (1989), expressed reservations about British membership of the ERM, pointing out that the EC's emphasis on budgetary discipline and price stability would lock Britain into a framework of deflationary policies which would undermine growth and threaten employment. This critique of the ERM reflected the substantial input of Bryan Gould who, as industry spokesman, was the convenor of the main economic policy review group. The party leader Kinnock and Labour's shadow chancellor, John Smith, rejected this position, however; they viewed ERM membership as an important part of Labour's commitment to a counter-inflationary policy and as a reassurance to the financial markets of the party's budgetary discipline. Labour's clear support for the ERM was settled following Gould's removal as industry spokesman in the autumn 1989.[19]

The Conservative government's decision to enter the ERM in October 1990 caused some difficulties for the Labour Party.[20] The straitjacket of an overvalued pound and high interest rates worsened the economic recession during Britain's membership of the ERM from October 1990 to September 1992. For the critics of the ERM in the party, this demonstrated the folly of getting locked into a fixed exchange-rate regime. The party's support for membership meant that its criticism of the Conservative government was blunted when the pound was forced out of the ERM in September 1992. Moreover, the exit from the ERM threatened to undermine a key plank in Labour's economic strategy. The party's response, however, was to reaffirm its support for the ERM provided it was complemented by a co-ordinated economic and industrial policy. At the same time, the Labour Party went further and supported the creation of a single currency (provided there was real economic convergence) as a means of countering the currency speculation which had provoked the ERM crisis.[21]

The party's support for the pound's entry into the ERM and, in principle, the single currency has been used by Labour to demonstrate its commitment to fiscal rectitude, counter-inflation policies and prudent economic management. The external economic disciplines of the EU would offer an incoming Labour government two additional advantages: first, a commitment to participate in European monetary integration would provide a useful safeguard against the sort of sterling crises and devaluations which had beset Labour governments in the 1960s and 1970s; and second, the monetary and fiscal constraints imposed by the EU could shield a Labour government from demands for an expansionary economic programme – 'the advantage of tying one's hands'.[22]

The development of the party's macro-economic policy in the 1990s has brought it in line with the economic objectives at the core of the EMU project. Under the leadership of Tony Blair, Keynesian demand management policies have been abandoned and the party has emphasised that price stability is the central objective of its economic policy.

> Macro-economic policy must be directed to keeping inflation low and as stable as possible.[23]

> Macro-economic policy must be kept tight, disciplined and geared to stability. I believe low inflation is the essential prerequisite of investment in the long term …[24]

The party's commitment to price stability was clearly demonstrated by its decision, within days of taking office in May 1997, to grant the Bank of England independence to set interest rates. This move also indicated that the Labour government could accept, in principle, an independent European Central Bank which will be an integral part of the institutional structure underpinning the single currency. The leadership of the Labour Party thus has few difficulties with the orthodox monetary and fiscal framework provided by EMU. Its main reservations are to do with the doubts surrounding the British economy's convergence with the other leading EU economies and the political difficulties the party is likely to encounter in attempting to get public support for the pound's entry into the single currency.[25]

The single currency issue is, however, potentially divisive for the Labour Party.[26] The remaining anti-EU elements in the party are predictably hostile to Britain's membership of a single currency. In addition, pro-Europeans on the left of the party are critical of the deflationary bias of EMU which could have damaging effects on growth and employment. The tensions in the party over the single currency issue were evident in the survey of Labour parliamentarians carried out in 1995–96:[27] 74 per cent of MPs disagreed with the statement that the establishment of a single currency would signal the end of the UK as a sovereign nation; only 16 per cent of MPs viewed EMU as undesirable; 29 per cent agreed that the single currency would institutionalise neo-liberal economic policy in Britain; 42 per cent agreed that Britain should never permit its monetary policy to be determined by an independent European central bank; and 78 per cent agreed that the UK should not seek to meet the EMU convergence criteria if the result would be increased unemployment.

What emerges clearly from this survey of Labour parliamentarians is a party which has little fundamental opposition to the principle of EMU but significant reservations about the fiscal and monetary orthodoxy inherent in the project and concerns about its impact on growth and unemployment.[28] While these reservations about the single currency are significant, it should be noted that parties of the left across the EU, most with strong pro-European credentials, have expressed similar concerns about the economic impact of the EMU project.

While the Labour government has ruled out the pound's entry into the single currency during the lifetime of the 1997 parliament, it none the less shares the German-style economic priorities central to the EMU project and government fiscal and monetary policies are likely to mean that Britain satisfies the convergence criteria to be eligible for EMU membership.[29] For the Keynesians in the party, however, the co-ordination of economic policies at the EU level should have job creation rather than low inflation as its primary objective. Such a 'Euro-Keynesianism' would require, however, a significant degree of economic co-ordination hitherto unseen at EU level. The Jospin government elected in France in May 1997 has attempted to secure an easing of the budgetary disciplines contained in the Maastricht convergence criteria and a shift in focus by the EU away from the overriding preoccupation with price stability towards measures to reduce unemployment. The French government has also called for a 'European economic government', based on Ecofin (the Council of Economic and Finance Ministers), to counter the influence and political independence of the proposed European Central Bank. The Blair government, however, has sided with Germany in its opposition to any dilution of the convergence criteria for qualification for the single currency.

THE TRADE UNIONS AND EUROPEAN INTEGRATION

The changing attitudes of British trade unions towards European integration have been an important factor in the Labour Party's conversion to a pro-European position.[30] In the late 1980s the trade unions, traditionally quite hostile to Britain's participation in the integration process, shifted to a more positive approach towards the EC. This conversion, in parallel with that in the Labour Party, reflected important changes in economic thinking on the British left and a response by the unions to policy developments in both the domestic and European arenas.

In the early 1980s, the official policy of the TUC supported British withdrawal from the EC. After 1983, however, the question of Britain's withdrawal was not debated by the TUC and 'attention shifted towards problems and opportunities which arose from membership'.[31] As in the Labour Party in the early 1980s, the trade unions' opposition to Europe was influenced significantly by the economic policies based on reflation and the restoration of national economic sovereignty at the heart of the AES. An important change in the thinking of the British left (including the trade union movement) led to the abandonment of the AES, with its focus on the nation-state and nationally based policies, and a recognition that effective economic strategies in an era of interdependence required co-ordination at a transnational level.[32] In this context, rather than an impediment to national economic policy, the EC was seen as providing an appropriate arena in which to develop co-ordinated economic programmes.

The TUC's response to the 'single market' programme indicated that it recognised both the constraints which the increasing globalisation of economic activities imposed on national policy choices and the opportunities afforded by co-operation at the European level.[33] The EC's development of a 'social dimension' to complement the single-market programme was an important factor in reinforcing the TUC's pro-European position. The trade unions' exclusion from national policy-making processes following the election of a Conservative government in 1979 encouraged them to look towards the EC as an alternative arena for influence over policy. The address by Jacques Delors, President of the European Commission, to the TUC in 1988 was important in widening support for a pro-European orientation in the trade-union movement.[34] The Delors' speech outlined proposals for the development of a 'social Europe' in which social rights would be guaranteed and workers and their representatives would have rights to representation on company boards. These EC policy initiatives were attractive to British trade unions which had seen a marked erosion of their rights and a significant diminution of their political influence during a decade of Conservative government. The Delors' proposals for a 'social dimension' to the single European market became entangled in British domestic politics, reinforcing the pro-European stance of both the TUC and the Labour Party, and sharpening the divide over Europe between the Conservatives and Labour.

Margaret Thatcher, the Conservative prime minister, rejected the social dimension advocated by Delors and supported by the social democratic and Christian democratic parties in the EC. She wanted to see the 1992 project limited to the creation of a free, deregulated European market, a position consistent with her domestic policy of 'rolling back the state' in the economic sphere. Thatcher's opposition to the social dimension was made clear in her address to the College of Europe at Bruges in September 1988:

> ... we certainly do not need new regulations which raise the cost of employment and make Europe's labour market less flexible and less competitive with overseas suppliers... And certainly we in Britain would fight attempts to introduce collectivism and corporatism at the European level... .[35]

The domestic dispute over the social dimension gave both the TUC and the Labour Party a pro-European cause which was potentially popular with the electorate and enabled them to portray Thatcher as an increasingly isolated figure in the EC.

On the issues of Britain's membership of the ERM and the single currency, the TUC has adopted pro-European positions largely in line with the Labour Party's stance. In 1989, the TUC favoured the pound's entry into the ERM provided co-operation on monetary policy was complemented by co-ordinated strategies for growth. The TUC has also supported Britain's participation in EMU even though it has expressed reservations about the timetable for the introduction of the single currency and fears that the deflationary bias in the convergence criteria might produce unemployment.[36] The divisions in the trade union movement on the EMU issue mirror those in the Labour Party.

Britain's membership of the EU is now widely accepted in the trade union movement, although differences remain over the scope and direction of political and economic integration. The TUC's conversion to a pro-European position in the post-1983 period was an important contributory factor in Labour's own parallel shift towards a positive stance on Europe. The trade unions' powerful institutional position within Labour's policy-making machinery provided important support for Kinnock's programmatic reform of the party, of which a more positive European policy was a central theme. The evolution of the TUC's pro-European position was not simply political opportunism based on a response to its domestic exclusion and the desire to see the return of a Labour government. European integration was embraced for the potential opportunities it offered to the trade union movement. This more positive approach to Europe was reinforced by the links established with other European trade unions as a result of working together in trans-European organisations such as the Economic and Social Committee (ECOSOC) and the European Trade Union Confederation (ETUC). In addition, sectoral affiliations between British unions and international trade union organisations operating at the European level were important in the development of a pro-Europe policy.[37]

PRO-EUROPEANISM IN THE LABOUR PARTY

Labour's shift to a pro-European stance finds support at all levels of the party. This is in marked contrast to those periods in the 1960s and 1970s when the party's official pro-European policy was opposed by significant elements of the PLP, the NEC, the trade unions and the party membership. Recent survey evidence indicates that the party's rank-and-file membership shares the positive Europeanism of official party policy.[38] The trade union movement too has embraced European integration since the latter half of the 1980s.

The party's conversion from hostility to European integration to general acceptance of Britain's membership of the EU is fully reflected in the PLP.[39] Traditionally the PLP has had majorities or significant minorities against British membership. A survey of Labour MPs in the 1992–97 parliament indicated, however, that the overwhelming majority favoured continued membership: only seven per cent of respondents agreed that Britain should withdraw from the EU.[40] Other findings from the survey indicated that the change in the party's stance on Europe reflected opinion in the PLP: for example, 59 per cent of respondents agreed that sovereignty could be pooled; 88 per cent agreed that the globalisation of economic activity made EU membership more necessary for Britain; and 52 per cent of MPs agreed that the benefits of membership for Britain have outweighed the disadvantages. The survey also found some support for Tony Blair's claim that,

> A younger generation of Labour MPs and activists is broadly supportive of Europe. The anti-Europeans tend to be older and less influential. Go to a Labour youth conference and attack Europe and delegates would look at you in blank incomprehension.[41]

MPs elected in the period 1950–83 were more sceptical than those elected in the period 1987–92. For example, no MP from the 1987–92 cohort would endorse Britain's withdrawal from the EU compared to 15 per cent of the 1950–83 cohort who would support it; 33 per cent of the 1950–83 cohort agreed that a single currency would signal the end of the UK as a sovereign nation compared to only 11 per cent of the 1987–92 cohort. In most respects, however, the findings indicated 'more scepticism across cohorts rather than within cohorts'.[42]

While there is overwhelming support in the PLP for continued membership of the EU, the survey also indicated that significant divisions remain in the PLP on some key European issues such as the single currency and the powers of the EU institutions. In the view of the authors, these divisions have the potential to create renewed conflict within the party over the issue of European integration.[43] Although the survey reveals clear divisions within the PLP on a series of European questions, many of the issues concerned are marginal or unlikely to become salient for the party. Much, of course, will depend on the progress of European integration and the challenges which the political and economic development of the EU poses for Britain.

While differences persist in the PLP over the scope, direction and pace of European integration, an insignificant number of Labour MPs is now opposed to Britain's membership of the EU. The transformation of the PLP into an essentially pro-European body is partly explained by the turnover of MPs; as we have seen, MPs first elected in the 1987–92 period tend to be less sceptical those first elected before 1983 and new Labour MPs, selected by local constituency parties, are likely to reflect the party's pro-European line. A 1996 survey of candidates who had either inherited safe Labour seats from a retiring MP or were contesting one of Labour's target seats indicated that 64 per cent thought Britain should join a single currency compared to 19 per cent opposed to such a move.[44] In addition to the parliamentary turnover, the emergence of a pro-European PLP is explained by individual conversions by previously anti-EC MPs and by the factional realignment which began in the party following Kinnock's election as leader in 1983.

> The so-called 'realignment of the left' set in motion a process of fragmentation as the old 'Bennite' coalition of 1979–83 splintered into two increasingly hostile sections, the hard and soft left. The soft left moved steadily towards a *rapprochement* with the leadership and their eventual inclusion into the governing coalition furnished Kinnock with overwhelming majorities in the PLP, the Shadow Cabinet and the NEC.[45]

With the support of the 'soft left', eager to see the party's return to office, Kinnock gradually eased Labour away from the commitment to withdraw from the EC towards a pro-European position. At the same time, the largely anti-EC 'hard left' of the party was increasingly marginalised in the party's decision-making structures and no leading party figures were identified with an anti-European position.

CONCLUSION

The Labour Party's shift from outright opposition to the EC in the early 1980s to enthusiastic support by the end of that decade for Britain's participation in the integration process represents one of the most significant recent changes in British party politics. The Conservative Party's increasing unease with the pace and direction of European integration since the late 1980s has allowed the Labour Party to usurp its position as the leading pro-European party. The Conservative Party's drift to a more sceptical stance on Europe opened a political space which a pro-European Labour Party was able to occupy. This reconfiguration of party alignments on the European issue has become more pronounced since William Hague became Conservative leader in June 1997 and quickly changed party policy to one of clear hostility to the single currency. There is little evidence to suggest, however, that a Conservative policy of disengagement from the EU will bring it electoral rewards; Labour's policy of 'constructive engagement' appears to be a more prudent electoral strategy and one which is likely to be supported by key business interests.

Labour's 'Europeanisation' has been a gradual transition, shaped by the interplay of domestic political developments, changes in party and trade union thinking, and the dynamics of Europe's economic and political integration. In terms of the domestic political context, Labour's embrace of Europeanism forms part of a broader strategy of programmatic and ideological renewal designed to make the party a credible alternative for government. At the same time, Labour has adjusted more readily than the Conservatives to the dynamics of the integration process since the latter half of the 1980s. Policy developments in the EU, such as the social dimension to the single market, have been attractive to the Labour Party and the trade union movement. After a long exclusion from national government, Europe offered a welcome alternative to the Conservative governments' economic and social policies. More importantly, the issue of national sovereignty, which was at the heart of Labour's traditional hostility to the European project and is the basis for Conservative divisions over Europe, has become a much less salient issue for the Labour Party in the 1990s. For example, the party's economic thinking has shifted away from the national protectionism of the early 1980s, to an approach which accepts the close interconnectedness of European economies and the EU as the appropriate context for co-ordinating member states' economic policies. The Labour Party's approach to the single currency clearly exemplifies this change in its economic policy and the declining importance of the sovereignty issue for the party. While the party remains cautious about development towards EMU, its concerns are primarily about the *economic impact* membership of a single currency would have on Britain rather than fears about the political and constitutional implications of joining the single currency.

While domestic political calculations, rooted in an electoral imperative, have been important determinants in Labour's more positive Europeanism, the change in the party's position is not mere political convenience. Labour's repositioning on the European issue reflects important changes in its policies and programme and an acknowledgement that Europe provides the most appropriate framework to pursue many of its policy objectives, particularly in the economic sphere. For Labour, pro-Europeanism became attractive, feasible and consistent with the party's policy goals. The pro-European stance finds support at all levels of the party and internal opposition to Britain's membership of the EU is now negligible. Divisions remain within the party over the scope, pace and direction of European integration. On an issue so wide-ranging, contentious and ever-changing this is to be expected. Nevertheless, the differences over Europe in the Labour Party are not comparable to the major fractures in the Conservative Party over the issue. The Labour leadership's tight grip on the party and policy, established under Kinnock, has been reinforced during Blair's tenure. This makes it highly improbable that an anti-EU position could make much headway within the party.

The Labour government which took office in May 1997 has pursued its declared policy of 'constructive engagement' with the EU. The new government quickly agreed to sign up to the Social Chapter and at the Amsterdam Summit in June 1997 it achieved its main objectives in the negotiations on treaty reform. The political mood in the EU was favourable to the incoming Labour government; the momentum towards political integration and far-reaching institutional reforms had stalled, however temporarily, and the British negotiators were not isolated on the key issues of institutional reform and policy development addressed by the summit. The Labour government was confident that its positive Europeanism, and the change in style and tone compared to the previous Conservative government, would enable it to shape the EU's agenda in a way which would advance British interests and make Europe a less contentious issue for the government.[46]

Nevertheless, the European issue remains potentially difficult for the

Labour government. The EU with which it has reached an accommodation is not static and further political and economic integration could reopen divisions in the party and cause domestic political difficulties. In particular, the issue of Britain's participation in a future single currency poses difficult and potentially divisive choices for the Labour government. In choosing to remain outside the first wave of single currency members, the government risks being isolated from the emerging 'hard core' of the EU and its influence diminished as a result. The European issue, the 'rogue elephant'[47] of British politics, retains its potential to divide parties and debilitate governments.

NOTES

1. 'Labour's aim is to work constructively with our EEC partners to promote economic expansion and combat unemployment. However, we will stand up for British interests within the European Community and seek to put an end to the abuses and scandals of the Common Agricultural Policy. We shall, like other member countries, reject EEC interference with our policy for national recovery and renewal.' *Britain Will Win* (Labour Manifesto 1987) in F.W.S. Craig (ed.) *British General Election Manifestos 1959–1987* (Aldershot: Parliamentary Research Services, Dartmouth 1990) p.473.
2. E. Shaw, *The Labour Party Since 1945* (Oxford: Blackwell 1996) p.181.
3. The 'Britain and the World' Group had Gerald Kaufman (shadow cabinet) and Tony Clarke (trade union) as joint convenors.
4. The reports were *Social Justice and Economic Efficiency*; *Meet the Challenge, Make the Change*; *Looking to the Future*; and *Opportunity Britain*.
5. The NEC ruled that the Policy Review documents could not be amended. On a card vote, the conference rejected two motions which had sought to give it the right to vote on individual sections of the policy review. See *Report of the Eighty-Eighth Annual Conference of the Labour Party 1989*, p.152.
6. Labour officially abstained from voting on the third reading of the bill in the Commons because it opposed the Conservative government's opt-out from the Social Chapter. On Labour's Maastricht tactics see K. Alderman, 'Legislating on Maastricht', *Contemporary Record* 7/3 (1993) pp.499–521.
7. 'There is a clear agenda for reform in Europe, including the European Commission, fundamental change of the CAP, enlargement of the EU to the East, completion of the Single Market, cooperation on infrastructure, science and technology. But there is a far better chance of achieving that reform through constructive engagement rather than mindless negativity.' From a speech to the Newscorp Leadership Conference, Hayman Island, Australia, 17 July 1995. Reproduced in T. Blair, *New Britain: My Vision of a Young Country* (London: Fourth Estate 1996) p.211.
8. Shaw (note 2) p.219. Kinnock describes his strategy as a 'two innings match'; see N. Kinnock, 'Reforming the Labour Party', *Contemporary Record* 8/3 (Winter 1994) p.543. On the policy and organisational changes in the party see P. Seyd, 'The Great Transformation', in A. King *et al. Britain At The Polls, 1992* (City, NJ: Chatham House 1993) pp.70–100.
9. Kinnock, ibid. p.539. For early doubts about the durability of Labour's anti-EC policy see M. Newman, *Socialism and European Unity: The Dilemma of the Left in Britain and France* (London: Junction Books 1983) pp.247–55.
10. Kinnock (note 8) p.542.
11. For a summary of British public opinion on the European issue see N. Nugent 'British Public

Opinion and the European Community', in S. George (ed.) *Britain and the European Community: The Politics of Semi-Detachment* (Oxford: Clarendon 1992) pp.172–201.

12. Labour's support for the Social Charter, officially announced in Dec. 1989, allowed the party to abandon its support for the closed shop since the Charter guarantees individuals the right not to join a union. Thus, the party was able to remove a contentious domestic issue from its agenda.

13. See *Wherever Next? The Future of Europe* (London: Fabian Society 1996) p.2.

14. An early indication of party's caution over Europe came in 1990 when the national leadership warned Labour MEPs that their support for a significant strengthening of the European Parliament went well beyond the party's position. See 'Labour MEPs Warned Off Federalism', *Guardian*, 16 March 1990.

15. Labour wants the EU to develop as an organisation based on independent nations co-operating in specific policy areas. It favours rapid completion of the Single European Market, enlargement, reform of the CAP and the common fisheries policy, an enhanced role for regional and local authorities in the EU, and greater openness and transparency in EU decision-making. In the run-up to the June 1997 Intergovernmental Conference it shared the Conservative Party's opposition to any erosion of the national veto on taxation matters, the budget, defence and security, immigration, and treaty reform. In contrast to the Conservative government, however, it supported the extension of qualified majority voting (QMV) to industrial, regional, environmental and social affairs, and a limited extension to the EP's powers of co-decision with the Council of Ministers. While the party pledged to sign up to the Social Chapter, it indicated that it would not support any proposals which would burden business and threaten jobs. See 'Brown says EU social costs would be vetoed', *The Times*, 11 Nov. 1996.

16. For example, in Blair's view 'there is not an insuperable constitutional barrier to a single currency, and there could in principle be benefits to it. But there has to be genuine and real economic convergence: *Guardian*, 4 Dec. 1995.

17. The threat which the single currency poses to parliamentary sovereignty and democratic decision-making is fundamental to the opposition to EMU voiced by a small number of Labour MPs: the most prominent exponents of this view are Tony Benn, on the left of the party, and Austin Mitchell on the right.

18. 'The next Labour government committed to radical, socialist policies for reviving the British economy, is bound to find continued membership a most serious obstacle to the fulfilment of those policies. In particular the rules of the Treaty of Rome are bound to conflict with our strategy for economic growth and full employment, our proposals on industrial policy and for increasing trade, and our need to restore exchange controls and to regulate direct overseas investment.' Reproduced in Craig (note 1) p.382.

19. Shaw (note 2) pp.184–5.

20. The party's immediate response was to indicate that it would be prepared, under the right economic conditions, to enter an EMU (see 'Labour Edges Towards EC Unity', *Independent*, 18 Oct. 1990). In Dec. 1990 an NEC statement confirmed the party's support for a single currency, based on firm convergence, and a politically accountable European central bank (see the reports in *Financial Times*, 14 and 18 Dec. 1990).

21. The new policy statement, *Europe: Our Economic Future*, was approved by 24 votes to two in the NEC: *Financial Times*, 24 Sept. 1992.

22. From F. Giavazzi and M. Pagano, 'The Advantage of Tying One's Hands', *European Economic Review* 32 (1988) pp.1055–82.

23. T. Blair, 'The British Experiment – An Analysis and an Alternative', The Mais Lecture, City University, London, 22 May 1995. Reproduced in Blair (note 7) p.82.

24. T. Blair, 'The Global Economy', speech to the Keidanren, Tokyo, 5 Jan. 1996. Reproduced in Blair (note 7) pp.121–22.

25. Chancellor Gordon Brown announced in Oct. 1997 that the pound would not join the single currency during the 1997 parliament because the five conditions he had laid down for Britain's economic convergence with its EU partners had not been met. The five economic

tests related to economic cycles, flexibility, investment, financial services and employment. For the Chancellor's statement to the Commons see *Financial Times*, 28 Oct. 1997, p.12.

26. For contrasting views on the issue from MPs see R. Berry, *Against a Single Currency* (London: Fabian Society 1995) Fabian Pamphlet 572 and K. Hill, *For a Single Currency* (Ibid. 1995) Fabian Pamphlet 573. A clear overview of the arguments is provided by Alan Donnelly, Labour MEP, 'Economic and Monetary Union', in G. Ford, G. Kinnock and A. McCarthy (eds.) *Changing States: A Labour Agenda for Europe* (London: Mandarin 1996) pp.1–16.

27. See D. Baker, A. Gamble, S. Ludlam and D. Seawright, 'Labour and Europe: A Survey of MPs and MEPs', *Political Quarterly* 67/4 (Oct.–Dec. 1996) pp.353–71.

28. This opposition was clearly expressed in the statement 'Europe Isn't Working' signed by 50 Labour MPs in March 1996. The statement called on Blair to rule out a single currency, warning that the budgetary criteria for EMU would require significant cuts in British jobs and public services: *Independent*, 29 March 1996. Forty-one of the 50 signatories were returned as MPs in the 1997 election.

29 Although Britain has an opt-out from the single currency, it is still required to meet the convergence criteria and its economic policy is subject to periodic review by the Council of Economic and Finance Ministers.

30. For full discussions of this change see B. Rosamund, 'National Labour Organizations and European Integration: British Trade Unions and "1992"', *Political Studies* 41/3 (Sept. 1993) pp.420–34 and D. MacShane, 'British Unions and Europe', in B. Pimlott and C. Cook (eds.) *Trade Unions and British Politics: The First 250 Years*, 2nd ed. (London: Longman 1991) pp.286–306.

31. Rosamund ibid. p.424.

32. Ibid. p.430.

33. As Rosamund has shown, affiliated unions differed in their response to the single-market programme. He distinguishes three groups: the 'pro-Commission group' whose policies were closest to the TUC strategy; the 'sectoral pragmatists' which were concerned mainly with the effects of '1992' on their own particular sectors; and the 'left sceptics' which regarded the single-market programme as one driven by the interests of multinational capital. Of these three factions, 'it was clearly the pro-Commission group which possessed the resources, voting strength and committee presence best suited to influence TUC policy on European integration'. (p.429).

34. The TUC's policy document *'1992', Maximising the Benefits, Minimising the Costs* (London: TUC 1988), was also presented to the 1988 congress. The congress approved a motion in line with the strategy outlined in the TUC report. See Rosamund (note 30) p.425.

35. Speech given by the prime minister on Europe in Bruges, 20 Sept. 1988. Office for the Minister of the Civil Service and the Central Office of Information, Dec. 1989.

36. At its 1996 annual conference, the TUC backed a General Council report favouring Britain's early entry into EMU; there was some dissent, however, with Unison abstaining on the issue and the TGWU opposing the single currency (see *Financial Times* 11 Sept. 1996). Bill Morris, general secretary of the TGWU, criticised parts of the trade union movement for being too eager to embrace the single currency and running ahead of the Labour Party on the issue (see B. Morris, 'Jobs on the line', *Guardian*, 9 Sept. 1996). The divisions resurfaced in Oct. 1997 when John Monks, general secretary of the TUC, called on the Labour government to take Britain into the first wave of single currency members (see 'Unions Split on EMU as Monks Urges Brown to Join First Wave', *Daily Telegraph* 22 Oct. 1997).

37. Rosamund (note 30) p.433.

38. 'Members were asked whether Britain should remain a member of the European Community or withdraw: overwhelmingly (89 per cent) they opt for remaining a member. Their commitment goes further: fewer than one in five (16 per cent) would resist moves to further integrate Britain into the European Community.' P. Seyd and P. Whiteley, *Labour's Grass Roots: The Politics of Party Membership* (Oxford: Clarendon 1992) p.47.

39. Labour's pro-European policy is also shared by the party's MEPs; see the survey evidence

reported in Baker *et al.* (note 27). With 62 MEPs, Labour is the largest national formation in the EP and provides the leader (Pauline Green) of the Party of European Socialists. The engagement with other European socialist parties has increased Labour's sense of influence in European political debate. For a discussion of the role Labour MEPs have played in the party's shift to a pro-European position see S. Tindale, 'Learning to Love the Market: Labour and the European Community', *Political Quarterly* 63/3 (July–Sept.1992) pp.289–90; S. George and D. Haythorne, 'The British Labour Party', in J. Gaffney (ed.) *Political Parties and the European Union* (London: Routledge 1996) pp.116–18; and D. Martin, 'Comment on "Assessing MEP influence on British EC Policy"', *Government and Opposition* 27/1 (1992) pp.23–6. On occasions, there has been a conflict between the national party leadership and MEPs on European issues. In May 1995, for example, it was reported that the national party put pressure on Labour MEPs to abstain on a resolution in the EP proposing the extension of QMV to policy areas where the national veto applied (see *Guardian*, 18 May 1995).

40. Baker *et al.* (note 27).
41. From a speech to the Royal Institute of International Affairs, London, 5 April 1995. Reproduced in Blair (note 7) p.287.
42. Baker *et al.* (note 27) p.369.
43. Ibid. p.370.
44. See the survey for 'A Week in Politics' (Channel 4) and *Observer*, 9 June 1996, p.4.
45. Shaw (note 2) p.219.
46. Shortly after Labour's 1997 election victory, Robin Cook, Foreign Secretary, declared his ambition to take Britain into a triangular leading role in the EU alongside France and Germany (see *Financial Times*, 8 May 1997). This view was reiterated by Blair in his interview in *Le Monde*, 7 Nov. 1997.
47. Description used by A. Mitchell and R. Heller, 'New Labour, New Deal', *New Statesman*, 26 July 1996, p.10.

Narratives of 'Thatcherism'

MARK BEVIR and R. A. W. RHODES

When confronted by the anarchy of political history, we domesticate it. The complex is made simple. The confusion of a myriad events is reduced to a chronology. We impose an order on discontinuity and change. It is inconceivable that people and events remain incomprehensible. This threat is removed by the stories – the narratives – we tell ourselves and one another about what happened. All too often, however, the search for understanding ends by reducing a complex multiplicity of narratives to a monolithic entity.

This contribution revisits the several accounts of 'Thatcherism' not to provide a comprehensive explanation of the phenomenon, but to show that the dominant traditions in British government provide distinctive narratives about it. There is no monolithic, unified notion of 'Thatcherism'. There is no heritage in the guise of a distinct policy programme or a political hegemony or a leadership style. The heritage of 'Thatcherism' lies in the dilemmas it helped to make salient in all the dominant traditions of British government, dilemmas which changed the ideas of each one. 'Thatcherism' is no more because it has been absorbed by the dominant traditions of British government which changed and continue to be changed as their exponents wrestle with and resolve the dilemmas. Although we speculate that future generations will see Thatcher herself as part of the Tory tradition about strong leaders, rather than the Liberal tradition about markets, our main concern is to highlight the multiplicity of stories about 'Thatcherism' and the continuing dilemmas created for the dominant traditions.

The next section explains the approach we have adopted, especially the notions of 'tradition' and 'narrative'. We then provide brief examples of several narratives of 'Thatcherism' embedded in various traditions to show there is no essentialist account. Finally, we show how 'Thatcherism' brought about change in British government, not because it was a unified phenomenon, but because it highlighted the salience of certain dilemmas in every tradition.

THE APPROACH: TRADITIONS AND NARRATIVES

It is rapidly becoming a commonplace that even simple objects are not given to us in pure perceptions but are constructed in part by the theories we hold true of the world. When we turn our attention to complex political objects, the notion that they are given to us as brute facts verges on the absurd. There is a sense, therefore, in which there is no 'Thatcherism' because all complex political objects are constructed in part by our prior theories of the world and the traditions of which they are part. We inherit a body of theories from our families, peer groups and the media. All our beliefs and actions emerge against the background of a tradition.[1] The tradition of theories we inherit provides the context in which we make sense of phenomena. How we understand 'Thatcherism' depends, therefore, on the theories within which we do so.

We define a tradition as a set of shared theories that people inherit and that form the background against which they construct the world about them. Traditions are contingent, constantly evolving, necessarily in a historical context. Traditions emerge out of specific instances and the relations between them. The instances that make up a tradition must have passed from generation to generation. They must encompass relationships as they develop over time with each case the starting point for later cases. Traditions must be composed of beliefs and practices relayed from teacher to pupil and so on. Moreover, because traditions are not fixed and static, we cannot identify or construct instances by comparing them with the key features of the tradition. Traditions are the product of the ways in which people develop ideas and practices. We can only identify any given tradition by tracing the appropriate historical connections back through time.[2]

Traditions consist of theories and narratives with associated practices. Theories are common to the natural and the human sciences. Narratives are the form theories take in the human sciences; narratives are to the human sciences what theories are to the natural sciences. However, studies of British government do not use the language of narratives. There is the related notion of an 'organising perspective' which provides 'a map of how things relate, a set of research questions'.[3] An organising perspective produces a narrative when we recognise it involves an implicit or an explicit historical story. The point we want to make by evoking narratives is, therefore, that understanding and explanation in the human sciences always take the form of a story. Narrative structures relate people and events to one another intelligibly over time, but these relations are not necessary ones. The human sciences do not offer us causal explanations that evoke physically necessary relationships between phenomena. Rather, they offer us stories about the past, present and possible futures; stories that relate beliefs, actions and institutions to one another by bringing the appropriate conditional and volitional connections to our attention. The human sciences rely on, therefore, the narrative structures found in works of fiction. However, the stories told by the human sciences are not fiction. The difference between the two lies not in the use of a narrative structure, but in the relationship of the narrative structures to our agreed understandings of what constitutes objective knowledge of the world. The human sciences provide narratives that relate those agreed understandings on objective knowledge about the world in a way fiction does not.

In this approach to the human sciences, it is possible to judge competing narratives by agreed standards of objectivity. Objectivity arises from criticising and comparing rival webs of interpretation in terms of agreed facts and established rules of intellectual honesty. The rules are accuracy and openness. Accuracy means using established standards of evidence and reason. So, we will prefer one narrative over another if it is more accurate, comprehensive and consistent. Openness means taking criticism seriously and preferring positive speculative theories which open new avenues of research and make new predictions supported by the facts. These rules provide the criteria for comparing webs of interpretation or narratives. The clear difference between this approach and conventional approaches to British government is that all interpretations are provisional. We cannot appeal to a logic of vindication or refutation. Objectivity rests on criteria of comparison. The web of interpretation we select will not be a web which reveals itself as a given truth. We will select the 'best' interpretation by a process of gradual comparison.[4]

So, a narrative provides a map, questions, a language and a historical story of British government. In what follows, we use our approach to consider 'Thatcherism'. Table 1 identifies four traditions and their associated narratives of 'Thatcherism'.

'THATCHERISMS'

A comprehensive historical review of each tradition would be unduly long. Our concern is to analyse 'Thatcherism' as it is constructed in the different British political traditions. So, we provide a summary of each tradition before discussing an example of its narratives of 'Thatcherism'.[5]

The Tory Tradition

The Tory tradition is elusive and relentlessly inconsistent.[6] All too often the Tory tradition is defined more by what it is not. Gilmour[7] argues the

TABLE 1
TRADITIONS, NARRATIVES AND 'THATCHERISMS'

Traditions	Tory	Liberal	Whig	Socialist
Stories	Preserving traditional authority	Restoring the markets undermined by state intervention	Evolutionary change	Role of the state in resolving the crises of capitalism
Narratives of 'Thatcherism'	Party and electoral survival	Reversing Britain's decline	Strong leadership and distinct ideology give new policy agenda	Failure of the developmental state
Examples	v.1 One nation (e.g. Gilmour 1992)	v.1 Markets (e.g. Willetts 1991)	v.1 End of consensus (e.g. Kavanagh 1990)	v.1 Political economy (e.g. Gamble 1988)
	v.2 Statecraft (e.g. Bulpitt 1986)	v.2 Culture (e.g. Skidelsky 1989)	v.2 Leadership (e.g. King 1985)	v.2 Developmental state (e.g. Marquand 1988)

Note: v. = version.

Conservative Party is not averse to change,[8] not a pressure group,[9] and not ideological.[10] More positively, 'the fundamental concern of Toryism is the preservation of the nation's unity, of the national institutions, of political and civil liberty'.[11] Blake argues Conservatives are against centralisation, equality and internal splits but, again to leaven the mix, for the national interest.[12] Gamble describes the British state as the Tory state with the defining characteristics of: racial and national superiority, a deferential attitude towards authority, secrecy surrounding the practice of high politics, an anti-egalitarian ethos and a status hierarchy.[13]

Some strands recur in the Tory tradition. For example, Michael Oakeshott provides the philosophical underpinnings for several raconteurs of Tory narratives. Ian Gilmour adopts Oakeshott's distinction between the state as a civil and an enterprise association.[14] An enterprise association is 'human beings joined in pursuing some common substantive interest, in seeking the satisfaction of some common want or in promoting some common substantive interest'. Persons in a civil association 'are not joined in any undertaking to promote a common interest ... but in recognition of non-instrumental rules indifferent to any interest'; that is, a set of common rules and a common government in pursuing their diverse purposes.[15] So a free society has 'no preconceived purpose, but finds its guide in a principle of continuity ... and in a principle of consensus'.[16] The Tory tradition favours civil association and only accepts the state as an enterprise association 'when individuals are able to contract out of it when it suits them'.[17] None the less Gilmour accepts that some state intervention will often be expedient, practical politics, essential to preserving the legitimacy of the state.[18]

For all its hedging about the role of the state, the Tory tradition upholds its authority. People are self-interested and hierarchy is necessary to keep order. Scruton makes the point forcefully: 'The state has the authority, the responsibility, and the despotism of parenthood.'[19] Strong leaders wield that authority to uphold national unity, correct social and economic ills and build popular consent.

We now examine briefly two narratives of 'Thatcherism'; as one-nation Toryism and as statecraft.

One-Nation Toryism. This narrative of 'Thatcherism' sees it as a threat to both the Conservative Party and to national unity. Gilmour is scathing about the 'dogma' of 'Thatcherism'. He argues that 'Thatcherism' is based on 'a simplistic view of human nature'. He disputes that 'everyone is driven by selfish motives' and that 'everyone pursues his selfish interests in a rational manner'.[20] Thatcher is not a 'true Conservative ruler' because she bullied people into conformity with her view of Britain as an enterprise association.[21] The economy was not transformed. Markets are not always right. 'The state cannot desert the economic front.'[22] 'Much social damage was also done'. 'British society became coarser and more selfish.'[23] His brand of 'one-nation Toryism' holds that if the state is not interested in its people, they have no reason to be interested in the state.[24] So, the government should '"conserve" the fabric of society and avoid the shocks of violent upheavals' and 'look to the contentment of all our fellow countrymen'.[25]

Statecraft. Bulpitt developed the Tory narrative of 'Thatcherism' as statecraft. He argues that 'what the Conservatives wanted to achieve in government was a relative autonomy for the centre ... on those matters which they defined as "high politics" at any particular time'.[26] So, the Conservative Party's main bias lies in its statecraft or 'the art of winning elections' or government survival. The main dimensions of statecraft are: a set of governing objectives; political support mechanisms able to build quiescent party relations and win elections; hegemony or winning the élite debate about 'political problems, policies and the general stance of government'; and a governing code or a set of coherent principles underlying policy related behaviour.[27]

The distinctiveness of 'Thatcherism' does not lie in its ideology or ideas but in its statecraft. Initially it focused on macro-economic management and foreign affairs, but showed no reluctance to extend its definition of 'high politics' and centralise when opposed. The details of Conservative policy are not the focus of analysis. The key point is that the Conservatives sought to achieve governing competence by redefining 'high politics' and increasing the centre's relative autonomy. This statecraft is a long-standing bias of the Conservative Party and 'there is a great similarity between the Conservative Party led by Mrs Thatcher and its predecessors under Churchill and Macmillan'.[28]

So, the Tory tradition in whichever narrative guise stresses tradition, authority and continuity, shares the story-line of party and electoral survival but produces divergent interpretations of 'Thatcherism'. Explaining the differences is easy. Gilmour stresses the break with tradition whereas Bulpitt stresses the continuities. Both share the concern with party survival and electoral success. They differ in their assessments of Thatcher's statecraft. Gilmour is a pessimist, seeing division and damage. Bulpitt is an admirer of Thatcher's ability 'to understand and work with the limitations placed on élite activity'.[29] All of which prompts an ironic conclusion. So often seen as a prime exponent of the Liberal tradition by both friend and foe, Thatcher typifies the Tory tradition with her commitment to strong leadership and her grasp of the arts of statecraft.

The Liberal Tradition

The narrative of 'Thatcherism' as the revival of nineteenth-century liberalism, with its faith in free markets, determined to slay the dragon of collectivism, and reverse Britain's decline, both economic and international, is one the clichés of British government in the late twentieth century. But like so many clichés, it did not become one without containing a large grain of truth. This narrative has its roots in the Liberal tradition's stories about markets and culture.

The Story about Markets. 'New Conservatism' revived the Liberal tradition by stressing freedom, applying the principles of freedom to the economy, and accepting the welfare state on sound Conservative grounds. Thus, Willetts[30] finds the roots of the 'New Conservatism' in the One Nation Group's arguments against government intervention and in such philosophers as Friedrich Hayek and Michael Oakeshott.[31]

For Willetts Adam Smith's 'system of natural liberty' provides the intellectual justification for free markets.[32] Markets tap 'two fundamental human instincts'; the instinct to better oneself and the instinct to exchange. These instincts, when 'protected by a legal order which ensures contracts are kept and property is respected' are 'the source of the wealth of nations'.

Big government cannot deliver prosperity, undermines markets and erodes communities. But 'Rampant individualism without the ties of duty, loyalty and affiliation is only checked by powerful and intrusive government.' So, Conservatism stands between collectivism and individualism and 'Conservative thought at its best conveys the mutual dependence between the community and the free market. Each is enriched by the other.'[33] The Conservative Party's achievement is to reconcile Toryism and individualism. It was also Thatcher's achievement.

'Thatcherism' is not the antithesis of conservatism because it too recognises there is more to life than free markets; it too sought to reconcile 'economic calculation with our moral obligations to our fellow citizens'.[34] Also its distinctiveness does not lie in 'Mrs Thatcher's actual political beliefs – very little of what she said could not have been found in a typical One-Nation Group pamphlet of the 1950s'.[35] It is distinctive because of Thatcher's 'political qualities'; her energy and conviction; her ability to move between general principles and the practical; and her judgement about which issues to fight.[36]

So, the 'Thatcherism' narrative in the Liberal tradition restores markets to their rightful place in Conservatism: it 'is within the mainstream of conservative philosophy'.[37] It also exemplifies great political skill. The government stuck to its principles and showed that the commitment to freedom meets people's aspirations and made them prosperous.[38] State intervention stultifies. Competition improves performance: '[f]ree markets are ... the route to prosperity'.[39]

The Story about Culture. The Liberal tradition also seeks to rescue Britain from economic decline and to restore the country's international standing. The origins of Britain's decline lie not only in state intervention undermining markets but also in cultural hostility to capitalism. So, the argument goes, the intellectual and political élite looked down on entrepreneurial behaviour, preferring 'the dream of New Jerusalem' in which there was no unemployment, poverty and ill health. But their illusion foundered on the 'British disease' of a poorly-educated and trained workforce, poor management, low industrial productivity and powerful trade unions. It was a dream which a bankrupted country could not afford and diverted scarce resources away from much needed industrial investment and reform.[40]

'Thatcherism' challenged the establishment. Its 'dream' is altogether different. It rejects the permissive society and the radical chic of the 1960s. It is a world of Victorian values – Samuel Smiles, not flower power, rules OK. It emphasises family values, self-reliance and the careful management of money. It dislikes trade unions and big government. It rails against dependence on the welfare state and praises the virtues of self-help and markets. It takes pride in Britain's imperial legacy and seeks to restore Britain's standing in the eyes of the world.[41]

Cultural change is a prominent strand linking the several essays in Skidelsky.[42] For example, Minogue argues that 'Thatcherism' renounces the culture of bourgeois guilt about the working class and minorities, rejecting 'the three fudging Cs' of caring, compassion and consensus for a culture of individual responsibility.[43] Whether this social Darwinism can reconcile the enterprise culture with social responsibility is open to debate. For example, Thatcher's call to Christian duty does not seem up to the task[44] and the electorate does not share Thatcher's values.[45]

So the narratives in the Liberal tradition stress markets and the affinity between markets and a culture of individual responsibility is obvious. Its story-line is to reverse Britain's economic decline through free markets sustained by an enterprise culture.

The Whig Tradition

The Westminster narrative fits well within the Whig tradition.[46] This narrative focuses on Britain as a unitary state characterised by: parliamentary sovereignty; strong cabinet government; accountability through elections; majority party control of the executive (that is, prime minister, cabinet and the civil service); elaborate conventions for the conduct of parliamentary business; institutionalised opposition, and the rules of debate.[47] It emphasises the historic heart of political science – the study of institutions or the rules, procedures and formal organisations of government; constitutional law; and constitutional history.[48] It speaks in the language of machine metaphors, employing such phrases as 'the machinery of government'.

The Whig tradition also has an idealist strand, seeing 'institutions as the expression of human purpose' and focusing, therefore, on the interaction between ideas and institutions.[49] It highlights 'how institutions and ideas react and co-operate with one another',[50] gradualism; and the capacity of British institutions to evolve and cope with crisis. Indeed, Whig historiography comes perilously close to telling the story of a single, unilinear, progressive idea, reason or spirit underlying the evolution of the British political system. Institutions provide the 'capacity for independent action, leadership and decision' while remaining 'flexible and responsive'. As important, the political science profession esteemed this tradition; they 'were largely sympathetic';[51] 'convinced that change needed to be evolutionary'; and celebrated 'the practical wisdom embodied in England's constitutional arrangements'.[52] The values of representative democracy, and the belief in the practical wisdom of the British Constitution still lie at the heart of the Westminster narrative.[53]

Although there is almost no discussion of power, the Whig tradition also makes some important, if implicit, assumptions. As Smith argues, it focuses on behaviour, motivations and individuals.[54] Power is an object which belongs to the prime minister, cabinet or civil service. So, 'power relationships are a zero-sum game where there is a winner and a loser' and power is 'ascribed to an institution or person and fixed to that person regardless of the issue or the context'. Personality is a key part of any explanation of an actor's power; personalisation. The Whig tradition's narratives of 'Thatcherism' contain these characteristics.

End of Consensus. Much of mainstream social science literature assesses the extent of change in British politics under the Thatcher government. The social sciences preferred method of working is to formulate hypotheses which can, in principle, be refuted or falsified. Gamble notes it 'introduced new rigour into British political science and widened the range of research questions but had no alternative narrative to propose'.[55] Behavioural methods were deployed, but the implicit historical story was that of the Whig tradition. So, although there is a vast literature, it differs less in its historical story-line and more in its subject matter.

Kavanagh uses the theme of 'the end of consensus', and an analysis of the interplay between events, ideas and actors, to argue the political agenda of British government has been substantially rewritten.[56] Consensus refers to agreement between political parties and governing elites about the substance of public policy'; the rules of the political game;' and the political style for resolving policy differences.[57] Thatcher had a distinctive set of new right-inspired policies: using monetary policy to contain inflation; reducing the public sector; freeing the labour market through trade union reform; and restoring the government's authority. These policies would free markets and create the enterprise society. He concludes the government was 'radical and successful';[58] 'reversed the direction of previous post-war administrations';[59] and that its policies, which appeared far-fetched in 1978, such as privatisation, are no longer exceptional.[60] In typical balanced, not to say Whig style, Kavanagh opines 'talk of permanent or irreversible changes may be too bold', but 'the Thatcher government has created a new agenda, one which a successor government will find difficult to reverse'.[61]

Riddell casts a more jaundiced eye over the Thatcher record, noting the often large disparity between ministerial rhetoric and achievements.[62] He considers 'Thatcherism' as 'essentially an instinct', not an ideology. He argues the Thatcher government provided a much needed shock to the British economy but it did not reverse the country's economic decline.[63] The political agenda has changed, the problems of decline are less acute, but there is a large legacy of problems – social division, crime, education and training, the welfare state and regulating the utilities.[64] These large areas of political controversy show the limits of the new consensus.

Leadership. King's elegant and influential essay focuses on Thatcher's style: 'the way in which she personally does the job'.[65] She is a 'very unusual' prime minister because she is a minority in her own party and she has a policy agenda.

> She ... had no choice, given her aims and determination, but to lead in an unusually forthright, and assertive manner. Partly this was a matter of her personality; she is a forthright and assertive person. But it was at least as much a matter of the objective situation in which she found herself. She was forced to behave like an outsider for the simple reason she was one.[66]

Assertive, however, does not mean she was cavalier: 'not only is she cautious, but she respects power and has an unusually well-developed capacity for weighing it'.[67] When confronted by a problem she does not think about organising the work, but about whom can help her: 'people, not organisations'.[68] So, she 'reaches out for decisions; she reaches our for people. She also reaches out for ideas.'[69] She is an actress, aware of her image and skilled in 'the presentation of self, the uses of self'. She is also 'a prodigious listener', 'a prodigious worker', 'a quick and eager learner' and 'considerate and solicitous' to those she trusts.[70] The political world is divided into goodies and baddies and deliberately, she uses fear as a weapon against the baddies.[71] King is aware of the weaknesses of Thatcher's style – the delays in decision making, *hubris*, the enemies she makes – but he concludes she has added to the repertoire of prime ministerial styles and made the job of prime minister bigger than before. In short, '[S]he is a formidable personality, and hers is a distinctive prime ministerial style'.[72]

Whatever the differences in their assessment of the Thatcher record, these narratives accommodate 'Thatcherism' to the Whig tradition in two ways. First, they identify the constraints on political action and the continuities in policy to domesticate the political convulsions of the 1980s. Thus, Kavanagh treats 'events' as a constraint on political leadership; recognises the changes had many causes; and muses how 'disappointment has been a fact of life for British ... governments'.[73] None the less there has been change and Thatcher is central to his explanation. So, second, these Whig narratives explain change by appeal to the personal power of

Thatcher. Kavanagh repeatedly describes her as the 'dominant figure'; and 'a remarkable figure'.[74] King stresses how she pushed out the frontiers of her authority.[75] Of course, 'We are not claiming that personal leadership is all important but Mrs Thatcher's personality and policies enabled her to take advantage of the constellation of events and ideas.'[76]

As ever, Riddell introduces a note of scepticism.[77] He stresses that the changes in macro-economic policy started in the mid-1970s which, coupled with international developments in the 1980s, would have led to 'shift away from collectivism towards individualism'. He dislikes the centralising and authoritarian tendencies of the government and is cautious about whether the 'success' of 'Thatcherism' depended on Thatcher. He prefers to describe her government as an example of a 'survivor regime', characterised less by its original policies and more by its determination to implement a rolling agenda of policies built out of its experience of government. More important, whatever the difference of emphasis, the story-line in these narratives assigns great explanatory power to Thatcher's personal qualities and her distinctive policies. Above all, they form part of the Whig tradition. Kavanagh makes the point succinctly: 'Over the long term continuity is more apparent than discontinuity.'[78]

The Socialist Tradition

The Socialist tradition, with its structural explanations focused on economic factors and class and its critique of capitalism, mounted a prominent challenge to Whig historiography. It disputes the factual accuracy of the Westminster narrative and challenges specific theoretical interpretations, although it is more likely to use the language of 'counterfactuals' than 'falsification' and 'refutation'. The historical story is anti-Whig. For example, Marquand comments:

> The old Whig historians were not wrong in thinking that Britain's peaceful passage to democracy owed much to the hazy compromises which unprobed ambiguities make possible. By the same token, however, once these compromises cease to be taken for granted ...arrangements of this sort are bound to run into trouble. ... Respect for the rules of the game will ebb away. ... In doing so, they have focused attention ... on the hidden presuppositions of club government itself ... And, as a result, these presuppositions have started to come apart at the seams.[79]

So, the Whig tradition collapses because it confronts a heterogeneous, pluralistic society in which authority has been demystified, cultural values have changed, the political system has lost legitimacy, and territorial politics

is in disarray.[80] However, the Whig tradition is still a common starting point
and it exerts a pervasive influence. Thus, Hall and Schwarz accept that
Britain has a unique political tradition characterised by stability and
continuity, although they stress crises and 'frenzied reconstruction' to
counter the focus of other commentators on continuities.[81] They still have to
recognise the 'passive transformation' of the UK; the marginalisation of
radical movements; the 'peculiarity of the British case'; the 'partial and
uneven' transition to collectivism; and the 'underlying persistence' of the
British political tradition.[82]

The Socialist narratives of 'Thatcherism' come in many guises with
many differences of emphasis. To show this variety, we provide brief
summaries of two influential accounts: Gamble's interpretation of
'Thatcherism' as the political economy of the free market and strong state;
and Marquand's account of the failure of the developmental state.

The Conservative Party did not invent the label 'Thatcherism', but stole
it from its radical parents. The social-democratic consensus of the post-war
period hid the contradictions of modern capitalism. That consensus began to
fall apart in the 1970s with Britain's continuing economic decline, world
recession, the failure of the Labour government and the revival of the Cold
War. The governing élite failed to restructure the British state and to
modernise the economy.[83] So, the Socialist tradition sees 'Thatcherism' as a
response to this crisis of capitalism.

The Political Economy of 'Thatcherism'. Gamble rejects all one-sided
explanations of 'Thatcherism' and uses political, economic and ideological
arguments to explain the fortunes of 'Thatcherism' in promoting both free
markets and a strong state.[84] He builds on the work of Hall who interprets
'Thatcherism' as replacing the existing social democratic ideology with its
own vision, creating 'a new historic bloc'.[85] However, this new ideological
discourse does not emerge, it is constructed. The Conservative hegemonic
project is 'authoritarian populism'. The term is deliberately contradictory to
capture the contrasting free-market and strong-state strands in
'Thatcherism'. The 'populism' encompasses: 'the resonant themes of
organic Toryism – nation, family, duty, authority, standards, traditionalism
– with the aggressive themes of a revived neo-liberalism – self-interest,
competitive individualism, anti-statism'.[86] The 'authoritarian' covers the
'intensification of state control over every sphere of economic life', 'decline
of the institutions of political democracy', and 'curtailment of ... "formal"
liberties'.[87] So the 1980s are characterised by, for example: centralisation,
the 'handbagging' of intermediate institutions, the refusal to consult with
interest groups, and state coercion. 'Thatcherism' stigmatises the enemy

within – for example, big unions, big government – while creating a new historic bloc from sections of the dominant and dominated.classes which seeks to establish hegemony.

The argument from politics sees 'Thatcherism' as one possible conservative, local response to the problems of economic recession and restructuring. Gamble concludes there is 'no Thatcherite electorate, no Thatcherite party, no Thatcherite consensus' but 'there have been several real and significant changes'.[88] The government did have a distinctive ideology and more than any other post-war government, it did have a strategy. It sought 'to build new coalitions of interest, to win the battle of ideas for a radical change of direction and the dismantling of old structures and old priorities'.[89] Its statecraft was not confined to winning office. The politics of 'Thatcherism' are part of a broader project to modernise British capitalism. Freeing the economy and strengthening the state were an attempt to create 'the basis for a new and viable regime of accumulation'. This argument from economics reasserted 'the traditional international orientation of British economic policy'. The government gave priority to 'the openness of the British economy over the protection of domestic industry'. This accumulation strategy sought to strengthen the integration of the British economy in the world economy.[90] It abandoned the manufacturing industries of the Fordist era for a free-market accumulation strategy which sought to make British companies competitive in the international economy. But for this strategy to succeed, the government also had to reform state and civil society. So, it asserted its authority by reducing government responsibilities, distancing itself from interest groups, and disciplining the public sector. The strong state would change attitudes and behaviour in civil society to support the free market.

Gamble is no apologist for 'Thatcherism'. He has a keen appreciation of its limits, recognising that it has not reversed economic decline; it was a transitional period. But he claims his multi-faceted account clears up the two central mysteries of 'Thatcherism'. 'Thatcherism' is not alien to Conservatism because its statecraft of restoring the state's authority and autonomy is part of the central Tory tradition. The gulf between ambition and achievement is explained by the scale of the task. The government knew where it wanted to go but its reforms of state and civil society were not up to that task.

The Developmental State. Marquand tries to answer two overlapping questions. Why did the Keynesian social democratic governing philosophy collapse? What are the main economic and political problems which a successor philosophy must address? He argues the collapse took place because Britain failed to become a developmental state. Britain failed 'to adapt to the waves of technological and institutional innovation sweeping through the world economy' and 'Britain's political authorities ... repeatedly failed to promote more adaptive behaviour.'[91] Britain failed to become an adaptive, developmental state because of a

> political culture suffused with the values and assumptions of whiggery, above all with the central Lockean assumption that individual property rights are antecedent to society. In such a culture, the whole notion of public power, standing apart from private interests, was bound to be alien. Yet without that notion, it is hard to see how a developmental state, with the capacity to form a view of the direction the economy ought to take, and the will and moral authority to put its views into practice, can come into existence.[92]

The Westminster narrative also inhibited an adaptive response. The basis of this narrative is parliamentary sovereignty which 'inhibits the open and explicit power sharing on which negotiated adjustment depends'.[93] The British crisis is a crisis of maladaptation coupled with: a loss of consent and growing distrust between governments and governed; possessive individualism or sectional interests dominating the common interest; and 'mechanical reform' or change through command, not persuasion.[94] In short, Britain failed to adapt because its political culture was rooted in reductionist individualism.

Although it is not his main concern, Marquand's account of 'Thatcherism' stresses the congruence between its market liberalism and British political culture of possessive individualism and the inability of both to deal with the crisis of maladaptation.[95] In short, the liberal solution deals with the consequences of state intervention, political overload and bureaucratic oversupply, not with the dynamics or causes of these processes. Possessive individualism is the cause of Britain's maladaptation, so it cannot provide the solution which lies in common, not individual, purposes and the developmental, not minimal, state. As a result, 'Thatcherism' contains three paradoxes.[96] First, the policies for a free economy conflict with the need for a strong, interventionist state to engineer the cultural change needed to sustain that free economy. Second, the wish to arrest national decline conflicts with the free trade imperatives of liberalism because of the weakness of the British economy. Third, the attack on intermediate institutions – the BBC, local government, the universities – undermines the Tory tradition which sees them as bastions of freedom; markets conflict with community.

So, the socialist narratives interpret the 'end of consensus' as part of the crisis of British capitalism stemming from its inability to become a developmental state. 'Thatcherism' is a local response to this crisis and is beset by internal contradictions. Free markets are a transitional solution for the open economy of a medium-sized industrial country operating in a global economy.

DILEMMAS AND THEIR EFFECTS

There is no essentialist account of 'Thatcherism'. Even the search for a multi-dimensional explanation is doomed. It is not a question of identifying the several political, economic and ideological variables and determining their relative importance. It is not a question of levels of analysis. It is more fundamental. The maps, questions and language of each narrative prefigure and encode different historical stories in distinctive ways. Historical stories as different as preserving traditional authority, restoring markets, gradualism and resolving the crises of capitalism construct the phenomenon of 'Thatcherism' in radically different ways. In deconstructing 'Thatcherism', we have shown there is no single notion to be explained. 'Thatcherism' as statecraft, as economic liberalism, as leadership and as a hegemonic project are different notions evoking different explanations. 'Thatcherism', then, was not an objective, given social phenomenon with a single clear identity, but rather several overlapping but different entities constructed within overlapping but different traditions.

What links the different constructions of 'Thatcherism' is not an agreement about the phenomena to be explained, but rather a recognition of the peculiar salience of certain dilemmas for British government since 1973. Proponents of all the traditions considered here understand 'Thatcherism' as a response to certain dilemmas and feel pushed by it to search for solutions to them. The legacy of 'Thatcherism' consists, therefore, not in a monolithic set of institutions, practices or beliefs, but rather in the diverse ways in which people inspired by different traditions have responded to these dilemmas.

A *dilemma* arises for an individual or institution when a new idea stands in opposition to an existing idea and so forces a reconsideration.[97] Four dilemmas occupy centre stage in our argument: welfare dependence, overload, inflation and globalisation. Although these dilemmas arise in all traditions, each tradition constructs the dilemmas differently and accords them different political salience. By introducing the topic, we court the danger of raising too many questions which cannot be answered in the compass of a single contribution. But it is important for our core argument and so, briefly, we discuss dilemmas. To recap, 'Thatcherism' as variously constructed highlighted the political salience of the dilemmas for all the traditions. It forced a reconsideration of existing beliefs. So, 'Thatcherism' lives on in the changes made in each tradition in response to the dilemmas. To discuss dilemmas is to focus on these continuing changes, on the diverse impact of 'Thatcherism', and on how each dilemma contributes to a discontinuity.

Welfare Dependency

The possessive individualism celebrated by the liberal tradition generally, as well as in its interpretation of 'Thatcherism', poses the dilemma of reconciling markets which deliver freedom and prosperity with community. For a conservative, commitment to community is 'the source of individual identity and satisfaction'.[98] We have moral obligations stemming from our social roles; we are born into duties. Free markets can destroy communities because they require the free movement of capital and labour. Willetts stresses the need to preserve community to distinguish his brand of conservatism from 'the economic liberal, without a trace of conservatism in him',[99] but sustaining community and Christian duty also involves meeting the spiralling costs of the welfare state. Care and compassion cost money. The Tory tradition's story about culture claims the welfare state creates welfare dependency, prompting calls for greater individual responsibility and a greater role for the family and self-help. One-nation Tories see the welfare state as a key way of conserving the fabric of society and the state's legitimacy. The invisible line between the 'nanny-state' and the caring state generates discontinuities as policy oscillates between penny pinching cuts, marketising services, privatising benefits and increasing public expenditure on social security and health to the highest levels in the post-war period.

Dependence is a dilemma common to all traditions. Thus, for New Labour, the welfare state now acts as a safety net more than as part of a search for greater social equality. The ideal of moral equality bequeathed to the party by ethical socialists no longer translates straight forwardly into rough economic equality. Instead, the traditional ethical socialist emphasis on the rights of the unemployed and disadvantaged is played down. The new story stresses that welfare recipients have duties to society. The socialist concepts of citizenship and welfare changed and continue to change.

Overload

For all its inadequacies, the overload thesis so popular in the 1970s drew attention to the limits to state authority.[100] One response noted in the narratives on 'Thatcherism' was to attack intermediate institutions such as the trade unions, local government and professional groups and reassert authority; to recreate the strong state. Several commentators note the market economy depends on the extensive use of state authority to bring about the changes in civil society necessary to sustain that economy, and yet such state intervention undermines markets.[101] Thus, the collectivist and libertarian strands in the British political traditions re-emerge in yet another guise. Too few commentators also note that state authority is itself undermined by markets or, to be precise, the marketisation of public services. Rhodes contrasts Britain as a unitary state with Britain as a differentiated polity and argues that interdependence, differentiation and functional power limit central authority.[102] As governments reform the state by creating markets and quasi-markets, they fragment bureaucracy, their main tool for exerting control. Policies are imposed. Implementing agencies rebel. The government depends on them to implement policies but too rarely seeks their compliance through negotiation. The attempt to exert authority, to act as the strong state, founders on the fragmentation and dependence created by state intervention, thus generating discontinuities as authority is flouted and intent and outcome diverge, often markedly.

Inflation

Inflation had become a major problem for the British economy by the end of the 1970s when the Labour government, under pressure from the International Monetary Fund (IMF), agreed to introduce strict monetary controls. The problem was often constructed through a monetarist critique of Keynesianism. During the 1980s and 1990s, proponents of all the dominant traditions came to accept four central tenets of monetarism: the key monetary levers should be interest rates rather than fiscal policy; the supply side of the economy should be considered more significant than demand management; low inflation should be as important a goal of economic policy as low unemployment; and government should develop monetary policy in accord with rule, not discretion, to preserve credibility. This change is central to many accounts of 'Thatcherism'. As important, the emergence of New Labour shows how socialists confronted the dilemma posed by monetarist ideas for their commitment to full employment and the welfare state. The Labour Party began to emphasise that economic recovery would bring unacceptable inflation unless it took place with a commitment to macro-economics stability and supply side policies to boost industry. Increasingly it opted for a more positive view of markets. Quasi-market mechanisms as well as privatisation are entrenched on the party's agenda. It rediscovered its notion of ethical equality and downgraded the emphasis on material equality. Here, in reworking the heart of the narrative, discontinuities were generated as the response to inflation distanced the Labour Party from both its legacy, and traditional sources of electoral and financial support. It cannot meet the call for redistributive policies from its traditional sources of support and the coalitions of new tribes, while refusing to increase both taxes and public expenditure.

Globalisation

We noted earlier the paradox in the Liberal tradition's narrative of Thatcherism between arresting national decline and the free-trade imperatives of liberalism. Reasserting national sovereignty is an illusion in a world where transnational power sharing is inevitable. So, the free-trade nostrums of liberalism undermine efforts to restore Britain's standing in the world and foreign policy is beset by the discontinuities generated by the clash of global markets and national sovereignty. Within the Socialist tradition, the dilemma is to transform socialism in one country into line with new economic patterns and new social groups. New technologies globalised key parts of the British economy and changed the interests and expectations of the working class. For example, the British economy could no longer go its own way but had to place itself in the heart of Europe. The European Union (EU) and its idea of a 'social Europe' was one way of defending and extending the classic welfare state.

CONCLUSIONS: CONSEQUENCES FOR BRITISH POLITICAL PARTIES

The history of the British political traditions cannot be written as the triumph of a collectivism, a fate bemoaned from Dicey onwards. Dilemmas fuel a process of adjustment which constructs and reconstructs the ideas in each tradition, leading to discontinuities. Rather it is best seen as a history of continuous conflict between evolving traditions. The effects of the dilemmas made salient in part by 'Thatcherism' on our understanding of British political traditions can be seen in the ups and downs of the political parties. We explore these effects under the headings of 'the revival of Toryism' and 'the end of history'.

The Revival of Toryism

The impact of libertarian ideas on the Conservative Party was great, but so was the damage, theoretical and practical. Free markets undermine community and yet the notions of tradition and community are central to the Tory tradition. Government policy bypassed and attacked intermediate institutions such as local government which historically were a key strength of the party. The 1997 electoral defeat corroded the justification that liberal ideas brought success. This defeat was caused in part by pervasive internal divisions focused not only on the EU but also on the differences between the Tory and Liberal traditions. All these reasons point to a revival of Toryism in the Conservative Party.

For example, Ferdinand Mount's radical critique contends that we have a 'shrivelled and corrupted understanding of the British constitution' which is leading to a 'degeneration of the institutions which are thus misunderstood'.[103] He wants to rescue our constitutional language from desuetude and to turn from constitutional pragmatism to an understanding of how history has shaped our constitution. He mounts his attack on complacency to restore civil association, providing an account of the British constitution which few would deny was conservative and rooted in a faith in historical evolution over rational, purposive intent. Above all he makes a clarion call reaffirming the Tory tradition. Even Willetts (1992) writing from firmly within the Liberal tradition recognises the force of the critique of markets from community, labouring long and hard to reconcile the two.[104] As one would expect, Gilmour argues Conservatives should 'readopt one-nation Toryism'.[105] The tension between Toryism and libertarian ideas is a recurrent dilemma in and for the Conservative Party and is most obvious in the party's conflicting stories of its recent history.

The End of History

We take this phrase from Fukuyama's, for it is possible to explore his questions without accepting his answers.[106] Has history as a single, coherent evolutionary ideological conflict ended? Are we witnessing the unquestioned triumph of liberal politics and economics? British socialism, confronted with the failure of the Keynesian social-democratic operating code, the monetarist critique and the rise of 'Thatcherism', reinvented itself as 'New Labour' with its shift from material to ethical equality, its commitment to free markets, and its notion of citizenship stressing duties, not just rights. The Labour Party sought the electoral comforts of the social-democratic middle ground. Even if the grand choice between liberalism and communism has been resolved in favour of liberalism, this grand, overarching ideological conflict has been replaced by a multitude of particularistic ideologies. The new tribes hold out the prospect of a cross-class radicalism, of new electoral coalitions. To claim the triumph of liberal politics is reductionist. The phrase clarifies little because it collapses parliamentary representative democracy with new forms of associational politics. The choice between varieties of liberalism is great and challenging. A single label will not do.

For example, Marquand asks, 'How can a fragmented society make itself whole?'[107] The answer lies in restoring the bonds of community. We must redefine our common purposes through the politics of mutual education and new forms of dialogue.[108] The details of Marquand's argument are not central here. Suffice it to note he appeals to both John Stuart Mill's notion of learning the arts of government by doing them, and to the collectivist strand in the British political tradition, with its emphasis on social justice and positive government. The notion of a British developmental state sustained by a new politics of mutual education, raises the prospect of a transformative socialism built on, but not synonymous with, the economics and politics of liberalism.

Fukuyama's end of history thesis domesticates history and reduces a complex, chaotic mess to a monolithic, unilinear story. In contrast, we have rejected the interpretation of 'Thatcherism' as another instance of the triumph of liberal politics. We avoid domesticating 'Thatcherism' and stress the complex and contradictory ways it was constructed within traditions which provide radically different narratives. 'Thatcherism' consisted of competing narratives of what it was. We are all familiar with TINA – 'there is no alternative'. Perhaps, however, a more pertinent palindrome would be TINT – 'there is no "Thatcherism"'. Whatever coherence this notion may have had, it has dissipated among the narratives of British government. Its impact continues to be felt by each tradition to differing degrees, but it does not form a single narrative. In place of the monolithic, unified phenomenon of British government textbooks, we offer a decentered, non-essentialist account. That said, 'the Queen is dead, long live the Queen' is an apt epitaph because 'Thatcherism' lives on. Its legacy is the diverse way people understood and responded to it. These responses produced distinctive narratives of 'Thatcherism' which highlighted certain dilemmas for the dominant traditions. If this argument seems to belittle the consequences of 'Thatcherism', that is not our purpose. By changing traditions, 'Thatcherism' changed both the practice of British government and our understanding of that practice.

NOTES

We would like to thank Lotte Jensen, Janice McMillan, David Marsh and Gerry Stoker for their helpful comments.

1. M. Bevir, 'The Individual and Society', *Political Studies* 44 (1966) pp.102–14; idem, *The Logic of the History of Ideas* (Cambridge: CUP 1998).
2. W.H. Greenleaf, *The British Political Tradition, Volume 1. The Rise of Collectivism* (London: Methuen 1983) provides a well-known analysis of the British political tradition (see also S. Beer, *Modern British Politics* (London: Faber 1965, 1982). He posits a dialectic

between two opposing tendencies: libertarianism and collectivism. Libertarianism stresses four things: the basic importance of the individual; the limited role of government; the dangers of concentrating power; and the rule of law (pp.15–20). Its antithesis, collectivism stresses: the public good; social justice; and the idea of positive government. Our view of tradition differs. We regard his opposing tendencies as ahistorical. Although they come into being in the nineteenth century, they remain static, acting as fixed categories, ideal types, into which he forces individual thinkers and texts, even different parts of the one text or different utterances by the one thinker. Instances cannot be constructed by comparison with the features of a tradition.

3. A Gamble, 'Theories of British Politics', *Political Studies* 38 (1990) p.405; Greenleaf (note 2) pp.3–8.
4. Our approach is not relativist but we do not have the space to develop the argument here. See: M. Bevir, 'Objectivity in History', *History and Theory* 33 (1994) pp.328–44; idem, 'Objectivity and its Other', ibid. 35 (1996) pp.391–401; idem (1998) (note 1); M. Bevir and R.A.W. Rhodes, 'Narratives of British Government' (forthcoming). R.A.W. Rhodes, *Understanding Governance* (Buckingham: Open UP 1997) Ch. 9.
5. We recognise there is no one-to-one correspondence between author and tradition. Some authors clearly draw on more than one tradition. Equally, the table and our examples are not comprehensive. We are illustrating an argument, not writing a review article on 'Thatcherism'.
6. For an incisive dissection see T. Honderich, *Conservatism* (Harmondsworth: Penguin 1991).
7. Gilmour, *Inside Right* (London: Quartet 1978) pp.121–43.
8. Ibid. p.121.
9. Ibid. p.130.
10. Ibid. p.132.
11. Ibid. p.143.
12. R. Blake, *The Conservative Party from Peel to Thatcher* (London: Fontana) Ch.11 and postscript.
13. A. Gamble, *The Free Economy and the Strong State* (London: Macmillan 1988) pp.170–71.
14. Gilmour (note 7) pp.92–100; idem, *Dancing with Dogma: Britain under Thatcherism* (London: Simon & Schuster 1992) pp.272–3.
15. Idem (note 7) p.98; F. Mount, *The British Constitution Now: Recovery or Decline?* (London: Mandarin 1993) pp.74–5; D. Willetts, *Modern Conservatism* (London: Penguin 1992) pp.72–3.
16. Gilmour (note 7) p.97.
17. Idem (note 14) p.272.
18. Idem (note 7) p.236.
19. R. Scruton, *The Meaning of Conservatism* (Harmondsworth: Penguin 1980) p.111. See also Gamble (note 13) p.170.
20. Gilmour (1992) (note 14) p.271.
21. Ibid. pp.271, 273.
22. Ibid. p.276.
23. Ibid. p.278.
24. Idem (note 7) p.118.
25. Idem (1992) (note 14) p.278.
26. J.G. Bulpitt, 'The Discipline of the New Democracy: Mrs Thatcher's Domestic Statecraft', *Political Studies* 34 (1986) p.26.
27. Ibid. pp.21–2; idem, 'Historical Politics: Macro, In-Time, Governing Regime Analysis', paper to the PSA Annual Conference, University of York, 18–20 April 1995.
28. Idem (note 26) p.39.
29. Ibid. p.39.
30. Willetts (note 15).

31. One Nation Group, *Change is Our Ally* (London: Conservative Political Centre 1954).
32. Ibid. Ch.6.
33. Ibid. p.182.
34. Ibid. p.47.
35. Ibid. p.52.
36. Ibid. pp.52–3.
37. Ibid. p.54.
38. Ibid. p.61.
39. Ibid. p.136.
40. C. Barnett, *The Audit of War* (London: Macmillan 1986).
41. See, for example, J. Gould and D. Anderson 'Thatcherism and British Society', in K. Minogue and M. Biddiss (eds.), *Thatcherism: Personality and Politics* (London: Macmillan 1987).
42. R. Skidelsky, 'Introduction', in R. Skidelsky (ed.), *Thatcherism* (Oxford: Blackwell 1989).
43. K. Minogue, 'The Emergence of the New Right', in ibid. pp.129–30, 141–2.
44. Skidelsky (note 41) p.22.
45. I. Crewe, 'Has the Electorate Become Thatcherite?', in ibid. p.44.
46. For a guide and references see Bevir and Rhodes (note 4) and L. Tivey, *Interpretations of Politics* (London: Harvester Wheatsheaf 1988).
47. Gamble (note 3) p.407. See also P. Weller, *First among Equals* (Sydney: Allen & Unwin 1985) p.16 and G. Wilson, 'The Westminster Model in Comparative Perspective', in I. Budge and D. McKay (eds.), *Developing Democracy* (London: Sage 1994) pp.190–93.
48. Greenleaf (note 2) pp.7–9; Rhodes (note 4) Ch.4.
49. Rhodes, ibid.; Gamble (note 3) p.409; N. Johnson, 'The Place of Institutions in the Study of Politics', *Political Studies* 23 (1975) pp.271–83.
50. Greenleaf (note 2) p.xi.
51. Gamble (note 3) p.411.
52. Ibid. p.409.
53. See, for example, P. Hennessy, *The Hidden Wiring* (London: Gollancz 1995); P. Norton, 'Constitutional Change', *Talking Politics* 9/1 (1996) pp.17–22.
54. M.J. Smith, 'Theoretical and Empirical Challenges to British Central Government', *Public Administration* 76 (Spring 1998).
55. Gamble (note 3) p.412.
56. D. Kavanagh, *Thatcherism and British Politics. The End of Consensus?* 2nd ed. (Oxford: OUP 1990).
57. Ibid. p.6.
58. Ibid. p.241.
59. Ibid. p.209.
60. Ibid. p.281.
61. Ibid. p.302.
62. P. Riddel, *The Thatcher Decade* (Oxford: Blackwell 1989). See also D. Marsh and R.A.W. Rhodes, *Implementing Thatcherite Policies* (Buckingham: Open UP 1992) (note 4).
63. Ibid. p.206.
64. Ibid. p.218.
65. A. King, 'Margaret Thatcher: the Style of a Prime Minister', in A. King (ed.) *The British Prime Minister*, 2nd ed. (London: Macmillan 1985).
66. Ibid. p.116.
67. Ibid. p.118.
68. Ibid. p.122.
69. Ibid. p.126.
70. Ibid. pp.130–31.
71. Ibid. p.132.
72. Ibid. p.133.
73. Kavanagh (note 56) pp.18, 238–41, 15.

74. Ibid. pp.243, 272, 276, 318.
75. King (note 65) p.137.
76. See also *inter alia* S.E. Finer, 'Thatcherism and British Political History', in K. Minogue (note 41); H. Young, *One of Us* (London: Macmillan 1989).
77. Riddel (note 62) pp.216–17.
78. Kavanagh (note 56) p.209.
79. D. Marquand, *The Unprincipled Society* (London: Cape 1988) p.198.
80. Ibid. pp.199–204.
81. S. Hall and B. Schwarz, 'State and Society, 1880–1930', in M. Langan and B. Schwarz (eds.), *Crises in the British State 1880–1930* (London: Hutchinson 1985) pp.8–12.
82. Ibid. pp.26–7.
83. T Nairn, *The Break Up of Britain* (London: Verso 1981, revised ed.).
84. Gamble (note 13), Ch.7.
85. S. Hall, 'The Great Moving Right Show', in S. Hall and M. Jacques (eds.) *The Politics of Thatcherism* (London: Lawrence & Wishart 1983) p.23.
86. Ibid. p.29.
87. Idem, 'Popular–Democratic versus Authoritarian Populism', in A. Hunt (ed.) *Marxism and Democracy* (London: Lawrence & Wishart 1980) p.161.
88. Gamble (note 13) p.222.
89. Ibid. p.223.
90. Ibid. p.225.
91. Marquand (note 79) p.145.
92. Ibid. p.154.
93. Ibid. p.176.
94. Ibid. pp.211–12.
95. Ibid. pp.72–81.
96. Ibid. pp.81–8, and D. marquand, 'The Paradoxes of Thatcherism' in Skidelsky (note 42).
97. Bevir and Rhodes (note 4). We acknowledge we do no more than introduce dilemmas in this article. For a more detailed discussion of the concept see Bevir (1998) (note 1).
98. Willetts (note 15) p.69.
99. Ibid. p.92.
100. A. King, 'Overload: Problems of Governing in the 1980s', *Political Studies* 23 (1975) pp.284–96.
101. Gamble (note 13); Hall (note 85); Marquand (note 79) and idem, 'The Paradoxes of Thatcherism', in Skidelsky (note 41).
102. R.A.W. Rhodes, *Beyond Westminster and Whitehall* 2nd ed. (London: Routledge 1988, 1992) p.407.
103. Mount (note 15) p.214.
104. Willetts (note 15).
105. Gilmour (note 14) p.276.
106. F. Fukuyama, *The End of History and the Last Man* (Harmondsworth: Penguin 1992).
107. Marquand (note 79) p.223.
108. Ibid. pp.231–2.

Institutions, Regulation, and Change: New Regulatory Agencies in the British Privatised Utilities

MARK THATCHER

National institutions are usually analysed as factors for stability in public policy. Institutions act as constraints in decision making: they constitute the framework within which policy is made and provide rules, norms and procedures that systematically distribute power amongst actors and structure their choices.[1] Institutional change is rare, being confined to crises[2] and/or it is slow, taking place incrementally.[3] As a result, stable national institutions give rise to enduring patterns of state roles, capacities and policy making.[4]

Britain is offered as a good example of the constraining effects of stable national institutions. It is prone to 'institutional inertia'[5] and its institutional peculiarities arise from very long-term factors, such as the timing and form of industrialisation, state traditions and the history of its bureaucratic development.[6] The British state is 'weak', unable to act decisively because of limited policy instruments[7] and constricting conceptions of its role, leading to the claim that Britain is even a 'stateless society'.[8] In the field of industrial and economic policy, the state has lacked the expertise, incentives and levers to formulate detailed plans and implement them.[9] Instead, the British policy style has been that of 'bureaucratic accommodation', with relatively closed policy communities making decisions through a series of compromises largely negotiated in private.[10]

The experience of regulation in the 1980s and 1990s, however, challenges the literature both on institutions and on policy making in Britain. Despite the claims for institutional stability, the formal institutional framework of regulation has been comprehensively modified. One of the most prominent reforms has been the development of new regulatory bodies, 'independent' of (at least organisationally) the central executive. Although independent regulatory bodies have a long history in Britain,[11] the 1980s and 1990s have seen major changes. A great variety of organisations has been established – from bodies with their own legal existence, powers, duties and funds determined by statute to advisory organisations created and funded by government departments. Almost all policy fields have experienced the development of new 'independent' regulatory organisations – from education[12] to the financial markets in the City of London.[13] The new organisations have frequently established themselves as major participants in regulation and policy making, with at least the appearance of great power.

The contrasts between general literature and regulation in Britain during the 1980s and 1990s give rise to theoretical and empirical paradoxes. At the theoretical level, the central paradox offered by Britain's experience of regulation is that institutions do not always act as constraints on change: they can also facilitate policy modification. Thus, institutional reform produces new policy dynamics, as new organisations develop their roles within the altered institutional framework. This results in policy change and influences the form of that change. Empirically, regulatory reform offers a series of paradoxes in the light of studies of the British state. The new regulatory bodies have developed expertise, have utilised their powers to the full and have become central actors in decision making, able to impose their views in the face of opposition by large powerful companies. As a result, there is discussion of Britain becoming a 'regulatory state'.[14] Yet, this development has been undertaken by Conservative governments with the declared intention of 'deregulation'. Moreover, the new bodies represent a fragmentation of power, as some of the powers of ministers have been transferred to the new regulatory bodies. But, the resulting more fragmented state is a more powerful one – a paradoxical development that calls for explanation.

Given the breadth of regulation,[15] the present discussion examines reforms of regulatory regimes in the privatised utilities and, in particular, the impacts of new economic regulators on these utilities. 'Regulatory regime' is defined here as the formal rules governing decision making, thereby covering the institutional, procedural and constitutional aspects of regulation, whilst the impacts considered concern policy making, the role and capacities of the state and policy styles in Britain. The privatised utilities analysed consist of telecommunications, gas, electricity and water. These serve as valuable cases for several reasons. First, they are economically and financially large and indeed strategic sectors.[16] Second, the institutional reforms introduced in the 1980s marked a clear break with the previous regulatory regime. Third, the new regulatory agencies have been very active; indeed they have been pioneers of change and have served as an example (to be followed or rejected) both in Britain and in other European countries.

This essay begins by briefly setting out the regulatory regime in the four utilities before the 1980s reforms. It then examines institutional reforms of the regulatory regime in the sectors, and, in particular, the position of the new independent regulatory bodies. Next, it looks at the activities of the regulatory bodies, analysing their role, and how and why they were able to play such a role. Finally, the implications of the case study are considered, both for views of the state and policy making in Britain, and for institutional analysis more generally.

UTILITY REGULATION FROM THE LATE 1940s TO THE EARLY 1980s[17]

In the late 1940s, the Labour government nationalised those utility suppliers which were not already publicly owned. Most suppliers were or became public corporations over the period 1947–84, although there were important variations in timing and organisational structure. The suppliers before privatisation were:

- Gas: British Gas from 1972;[18]

- Electricity: in England and Wales, the Central Electricity Generating Board (for generation and transmission) and 12 area Boards (for transmission) from 1957;[19]

- Water: from 1973, in England and Wales, ten water authorities were established, which were also responsible for sewerage.[20]

- Telecommunications: the Post Office before 1981 and BT (British Telecommunications) from 1981.[21]

The regulatory regime was largely statutory.[22] Legislation gave most powers to the suppliers or ministers.[23] Parliament had few direct powers and the consumer councils created to represent users had no powers. Independent regulatory bodies were not created.

The regulatory regime offered a division of roles among the different actors. In particular, the suppliers were to enjoy operational autonomy, having an 'arms-length' relationship with ministers. They were given a monopoly over the supply of utility services[24] but were to use it in 'the public interest'. Thus, the suppliers, headed by Boards, had a legal identity and had powers over staffing and expenditure. Their statutory duties were broad. They generally included some form of overarching 'public interest' duty, notably to 'exercise' their powers to meet the social, industrial and commercial needs of Britain' and a general universal service obligation such as 'satisfying demand throughout Britain except if impracticable or not reasonably practicable'. In addition, however, the suppliers were to operate 'efficiently' and were expected to break even 'taking one year with another'.

Ministers were to deal with general 'policy' matters. The relevant minister(s) appointed Board members (but could not dismiss them other than in exceptional circumstances, such as incapacity). They could give 'directions' to the Board (mostly 'general directions') and could demand information. In financial matters, notably capital investment plans and borrowing, approval by the minister and sometimes also the Treasury, was required.

The operation of the regulatory regime in practice was a 'game' largely played between the utility suppliers, large manufacturers firms that depended on utility orders, and ministers and their civil servants. The 'game' was highly closed; occasionally outsiders such as trade unions penetrated it, but even so, mostly on employment-related matters. Users played little role. The decision-making processes were informal and involved discussions and negotiations conducted in private.

Given the participants in decision making, it is unsurprising that the interests of the utility suppliers and manufacturers were at the forefront of policy. The interests of users were, it was held, best served by technological developments led by suppliers, and also by prices that involved cross-subsidisation between services and hence between groups of users; in particular, businesses tended to cross-subsidise domestic users. Almost no attention was given to competition: the utilities were regarded as 'natural monopolies'.

Serious problems arose in the regulation of the nationalised industries, as the different actors did not play the roles that had been envisaged.[25] Ministers increasingly intervened in 'operational' matters such as price setting, employment and wages, often in pursuit of inflation, employment and regional development policies. Moreover, ministers, and especially the Treasury, limited investment by the nationalised industries according to the short-term demands of public-sector borrowing requirements. However, the suppliers held a virtual monopoly over information and expertise, especially in highly technical matters; hence they were able to implement their desired programmes, such as digitalisation of switching and nuclear power. Ministerial control was focused on short-term aggregate financial implications and no overall conceptual framework for intervention was developed. Indeed, the aims and priorities of the suppliers and of ministers were unclear and largely undefined, as social objectives such as universal service and cross-subsidisation of certain categories of user and services jostled with required rates of return, encouragement of exports by manufacturers, protection of employment and pursuit of government inflation and borrowing targets. Moreover, policies were rarely consistent as one set of short-term pressures and constraints succeeded another.

INSTITUTIONAL REFORM: THE NEW REGULATORY REGIME FOR THE
PRIVATISED UTILITIES

The 1980s and early 1990s saw the privatisation of the publicly owned
utility suppliers, driven by dissatisfaction with the previous regime, fiscal
pressures and the Conservative governments' view of private ownership
and party political advantage.[26] The sale of the utilities was not planned, but
began with the sale of just over 50 per cent of shares in BT in 1984 and
ended with the sale of British Energy in 1996. By 1997, only the Scottish
water and sewerage companies and the Northern Ireland Water Executive
were still publicly owned.[27] Moreover, the statutory monopolies of the
suppliers were removed, and the suppliers themselves (and any new
entrants) were given licences under which they were to operate.[28]

Most attention in the 1980s was focused on the principle and form of
privatisation. Yet privatisation also saw the establishment of new regulatory
regimes for the four utilities, in large measure through statutes. The main
statutes were:

- the Telecommunications Act 1984;
- the Gas Act 1986;
- the Electricity Act 1989;
- the Water Act 1989;
- the Water Industry Act 1991;
- the Water Resources Act 1991;
- the Competition and Service (Utilities) Act 1992;
- the Environment Act 1995;
- the Gas Act 1995.

The regimes for the four utilities had much in common: many of the
regulatory provisions (and indeed, often their exact wording) were copied
from the 1984 Telecommunications Act in subsequent legislation. Thus, one
can speak of 'the' regime for the four industries.

In each industry, a new regulator or 'Director General' (referred to here
as the DG or industry DG) was established by the privatising statute with
functions and powers in economic regulation. Each DG is the head of an
Office that has became known by its acronym. The industry DGs and their
offices are:

- Telecommunications: the Director General of Telecommunications
  (DGT) heading *Oftel* (the Office of Telecommunications);

- Gas: the Director General of Gas Supply (DGGS), heading *Ofgas* (the
  Office of Gas Supply);

- Electricity: the Director General of Electricity Supply (DGES), heading *Offer* (the Office of Electricity Regulation);

- Water: the Director General of Water Services (DGWS) heading *Ofwat* (the Office of Water Services).[29]

The new regulators were not the centre of political debate at the time of privatisation. Indeed, in 1984, the establishment of the Director General of Telecommunications was in part an attempt to counterbalance the failure to break up BT, and was seen as an interim measure, since it was expected that regulation would wither away and become unnecessary as competition was extended.[30] Expectations in the 1980s and the lack of an overall vision of the future position of the DGs may help explain the limited detail governing the activities, role and power of the DGs under the statutes. Those expectations also provide an important part of the explanation of the paradox of a 'deregulating' government creating powerful regulators, together with the decisions to give priority to securing rapid sales, support for privatisation from managers, large immediate revenues and 'windfall' gains for purchasers rather than the more difficult, dangerous and lengthy process of structural reforms to maximise competition.

Six features can be highlighted in the analysis of the DGs' place in the regulatory regime. First, the DGs were given important functions and powers. But, second, their duties, functions and purposes were very broadly defined. Third, in many areas, the statutes were largely silent. Moreover, a fourth feature is that considerable potential existed for tensions and contradictions among the different powers, aims and roles of the DGs. Fifth, the specific powers of the DGs were limited in number. Finally, individuals and bodies other than the industry DGs were given important powers and responsibilities, sometimes over the DGs or in domains in which the DGs had functions. These features can be seen in a brief analysis of the appointment of the DGs, their duties, their functions and powers and the statutory rules governing their decision-making procedures.

The industry DGs are appointed by the relevant secretary of state[31] for renewable terms of up to five years. The statutes do not specify the mode of selection, nor the criteria for choices; they merely exclude MPs. Dismissal of a DG is difficult – the legislation only permits it in circumstances of 'incapacity or misbehaviour'. The choice of individual is, at least in legal terms, crucial, since powers, duties and functions are vested in the DGs as individuals and not in their offices.

The statutory duties of the DGs in effect serve as their purposes. They are very broadly defined and lack detail. Moreover, legislation states that each DG should exercise his/her powers 'in the manner that he considers best calculated' to fulfil the various duties. The room for discretion is increased by the potential for the different duties to conflict with each other.

The statutes lay two primary duties on the DGT and DGWS, and three on the DGES and DGGS:

- to ensure that supply of services meets 'all reasonable demand' or all demand in so far as practicable;[32]

- to ensure that suppliers are able to finance the provision of such services;[33]

- in electricity and gas, to promote/secure competition.

In addition to these, there are secondary duties. These are numerous and vary somewhat between the four regulators, but generally include:

- promoting or protecting consumer interests;

- maintaining/promoting/facilitating effective competition (if not already a primary duty);

- promoting research and development;

- ensuring safety;

- protecting the environment;

- protecting certain categories of service (for instance, emergency services);

- protecting specific types of user (for instance, rural dwellers, the elderly, disabled and chronically sick).

With respect to the procedures to be followed by the DGs, the statutes cover specific types of decisions and situations. Thus, for example, in decisions over the modification and enforcement of licence conditions, they impose requirements on the DGs to publish Notices, and to consult, with minimum time periods for responses, to consider all non-frivolous representations and to provide reasons. In making decisions about performance standards, the DGs must consult licensees and a representative sample of persons affected. Public registers of Notices, licences and modification and enforcement orders must be kept. The DGES, DGWS and DGGS must consider matters referred to them by the user committees for electricity and water, and the statutory Gas Consumers Council. Moreover, the DGs must provide information via an annual report to the secretary of state that is then placed before parliament. But, aside from the specific types of decision mentioned by the statutes, there are few general statutory provisions governing decision making. Thus, for instance, there is no general statutory duty to give reasons or to consult.[34] Moreover, the DGs do not have a general duty to publish information. The statutes do not specify the methods to be followed by the DGs in determining their overall approach to regulation.

Many of the statutory provisions concern licences, since these are placed at the heart of the new regulatory regime (most suppliers are obliged to operate under a licence). A complex division of powers and functions is established by the statutes. On the one hand, licences are issued by the relevant secretary of state; the DGs can only issue licences with the authorisation or consent of the secretary of state. The exception is gas, where the 1995 Gas Act authorised the DGGS to issue licences (but the secretary of state determines the standard terms for licences). Moreover, the statutes provide very few rules as to the distribution of licences or the determination of the terms of licences. On the other hand, enforcement of licence conditions is almost solely a matter for the DGs, who are given powers to demand information relating to licence enforcement and to issue enforcement orders (either provisional or final) to licensees.[35] As noted, the legislation lays down detailed procedural rules for licence enforcement.

Licence modification sees powers given to several actors. Modification of conditions can occur via two routes. First, an industry DG and licensee can agree a change; however, the secretary of state has the power to block the alteration. Second, the industry DGs can refer any matters relating to the functions of licensees to the Monopolies and Mergers Commission (MMC), asking it to investigate and report on whether specified matters operate, or may be expected to operate, against the 'public interest' and if so, whether the adverse effects could be 'remedied or prevented' by a licence modification. If the MMC issues a report favourable to licence changes,[36] then the DG must alter the licence, 'having regard to' the MMC's report. In so doing, he/she can impose the licence modifications on the licensee; ministers have few powers to intervene.[37] The DGs enjoy great discretion over whether to make references to the MMC, and some (at present contested) discretion over the extent to which they must follow MMC report recommendations.[38]

Whilst the DGs' statutory functions and powers centre on licences, they have been given other, specific tasks. The DGs play an important part in the standards of performance to be achieved by licensees, especially after the Competition and Service (Utilities) Act 1992. Thus, for instance, they have been empowered to set performance standards, to collect and publish information on the achievement of such standards, to order modifications in the complaints procedures of licensees, to establish compensation for breaches of performance standards in individual cases and to determine disputes about matters such as undue discrimination and deposits.

More general statutory functions for the DGs do exist, but they are not accompanied by autonomous powers. Thus, for instance, the DGs are to advise the secretary of state, but are given few powers over the latter's decisions. They are to collect information on their sectors, but are not given a general power to demand information. The DGs can exercise concurrently certain of the Director General of Fair Trading's (DGFT) powers in their industries, notably to deal with courses of conduct detrimental to the interests of the consumer and monopoly situations under the Fair Trading Act 1973 and anti-competitive practices under the Competition Act 1980. However, under general competition law, many of these powers are to refer matters to the MMC; many decisions lie with the secretary of state, who can order the DGFT/industry DG not to proceed with an investigation, can make a reference to the MMC himself/herself and, if the MMC report finds breaches of competition law, can issue orders.[39] The industry DGs are not given major powers or duties under the statutes concerning takeovers and mergers.[40]

Thus, the position of the industry DGs under the various statutes is ambiguous: while given functions and powers, the significance and power of the DGs relative to other regulators, notably the secretary of state and competition authorities, and over the licensees, was far from clearly determined in the regulatory regime for the privatised utilities.

THE INDUSTRY REGULATORS IN ACTION

The industry DGs have been able to establish themselves at the heart of utility regulation, using their institutionally derived tools and taking advantage of the breadth and flexibility of the post-privatisation institutional framework of regulation. They have developed an extended sphere of activity, have participated vigourously in all the major regulatory questions in their industries (often attracting attention and controversy) and have accumulated experience, expertise and power. The development of the position of the DGs was not foreseen at the time of their creation, and has been far from easy. The DGs have faced pressures arising from conflicting demands: on the one hand, dissatisfaction with the events after privatisation (especially price rises for certain services) has led to calls for the DGs to remedy these substantive complaints; on the other hand, the appropriateness of unelected officials dealing with 'policy' issues has been questioned.[41] Nevertheless, the DGs have developed strategies to deal with such pressures; whilst differences have emerged, both over time and among the individual regulators, certain key common features and patterns can be discerned in their approach and roles.

Conceptual Frameworks: Competition and the Interests of the Consumer

The industry DGs have developed a conceptual framework concerning the purpose of regulation that has guided and justified their decisions. From their long and varied list of duties, they have given pride of place to promoting and securing effective competition and to protecting the consumer.[42] The regulators have claimed that the two are not only compatible, but also that the first is the best method of achieving the second.[43] They have given their other statutory duties of much less prominence.

The DGs have, therefore, pressed for the extension of competition. But, competition has not meant the end of regulatory action; on the contrary, the DGs have argued that to promote 'fair and effective' competition and to protect consumer interests, active regulation is needed, often to restrict the behaviour of the dominant former monopolist. In particular, they have used the concept of 'incentive regulation', whereby regulatory measures create incentives that 'mimic' those in a competitive market, an aim given particular importance when effective competition was not possible.[44]

The DGs' conceptual framework has provided them with a message and the ability to influence the agenda of regulation, notably concerning the extension of competition (discussed below). Moreover, it has offered the DGs great freedom of action, because of its high level of generality, the breadth of measures that can be justified as required to ensure 'fair and effective competition' and 'protection of the consumer' and its flexibility, so that the DGs can accommodate their other statutory duties. Thus, in addition to measures directly linked to extending 'fair and effective' competition, the DGs have pursued aims such as preventing suppliers from earning 'excessive' rates of return, encouraging cost-based pricing, supporting measures that reduce suppliers' costs and introducing quality standards, ensuring sufficient investment, fulfilling 'social obligations' and promoting 'public interest' measures (see below). The breadth and flexibility of the DGs' conceptual framework has helped them to justify action in response to public dissatisfaction with the privatised utility suppliers.

Visibility, Personalisation and Independence

The DGs have enjoyed considerable public attention and prominence. Individual DGs such as Sir Bryan Carsberg, Don Cruickshank, Clare Spottiswoode and Stephen Littlechild have become well-known figures.[45]

Their pronouncements, decisions and individual characteristics have been the subject of discussion and controversy not only in closed industry circles, but also elsewhere, including parliament and the mass media. The disagreements and sometimes sharp conflicts that have arisen between the industry DGs and the former monopolies (notably between Sir James McKinnon and Clare Spottiswoode, and British Gas, and between Sir Bryan Carsberg and Don Cruickshank, and BT) have been portrayed as almost 'David and Goliath' struggles between individual DGs and large powerful corporations. Moreover, the existence of individual regulators, as opposed to Boards, has strengthened the impression of powerful personalities.

The importance of the industry DGs has also rested on their apparent independence. The extent of their autonomy in decision making in reality is more difficult to gauge. There is little public information on the extent of direct intervention by ministers or their civil servants in the DGs' exercise of their powers. However, the DGs have spoken out, sometimes openly opposing government policy; one example is the 'windfall tax' imposed by the new Labour government in 1997, which was publicly queried by Byatt and Spottiswoode.[46] Moreover, appointments have not been party partisan: although the DGs have been sympathetic to government policy of extending competition among private suppliers, they have also been strong-minded capable individuals with expertise rather than mere party supporters.[47]

On the other hand, informal contacts between the DGs and the executive have been common, and ministers have not only openly commented on the DGs' decisions, but, at times, have also sought to intervene in the DGs' spheres. This has been particularly apparent with the change of government in the 1997 general election. Thus, for instance, John Battle, the new minister at the DTI, told Littlechild and Spottiswoode to review the operation of their markets, emphasised the importance of social obligations and commented on the way in which competition in domestic markets would be expected to operate.[48] Ministers also acted to convene the 'water summit' of May 1997 to plan to reduce leakages;[49] not only had Ofwat been working on leakage targets for some time, but mandatory targets would involve a licence change. Price reviews in water, in which there have been conflicts between Ofwat and the National Rivers Authority (NRA) over environmental obligations, costs and price caps, have seen the active involvement of the government: 'Quadripartite machinery', with participants from Ofwat, the NRA, the Department of the Environment and the water companies, was set up to examine the issues[50] and Ofwat even stated that prices (which are licence conditions and hence part of Ofwat's sphere) are 'ultimately' a matter for the secretary of state.[51] Finally, although no DG has yet been dismissed during his/her term of office, Sir James

McKinnon was not reappointed as DGGS in 1993, which may have been related to his highly conflictual relations with British Gas, while the prospects of certain DGs criticised by Labour provoked speculation in 1997.[52]

The DGs and Decision-Making Processes in Utility Regulation

The industry DGs have gained a central role in general policy making. Their offices have developed great expertise and a wealth of information, on which government departments are dependent. In major policy decisions, the DGs are consulted extensively, and indeed are often active participants; this is seen most clearly in licensing and also in mergers and takeovers (see below). Moreover, the DGs have developed links with parliament: they have provided evidence and appeared before select committees (notably the Trade and Industry Select Committee, and also the Public Accounts Committee),[53] in addition to less formal contacts with MPs and peers.[54]

The power of unelected DGs has left them open to criticism.[55] But, they have developed a series of procedures for decision making above and beyond the statutory requirements placed on them, that increase the acceptability of their role and provide a defence for their lack of procedural legitimacy and, in particular, being unelected officials making policy. The DGs have claimed that the procedures provide 'transparency' and opportunities to participate in decision making for all interested parties; they certainly represent somewhat increased formality and openness compared with policy making before privatisation. Thus, almost all major decisions (usually taking the form of licence changes), such as price reviews, interconnection arrangements or quality of service standards, begin with a consultative document being issued. The DGs then allow a period for representations, before issuing definite proposals, and then seeking to introduce these. Since 1995, Oftel has held oral hearings and debates.[56] Informal 'advisory panels' have also been appointed by DGs, to offer advice and to counterbalance the charge of over-individualised decision making. The DGs' offices publish large quantities of information – far greater than that produced before privatisation – on matters ranging from the evolution of their industries to quality of service and complaints. In addition, the DGs have claimed to listen carefully to the voice of the consumer, not only through consultation, but also via market surveys and bodies established to represent consumers in each industry, such as the advisory bodies in telecommunications,[57] Customer Service Committees in water and area Consumer Committees in electricity, and in gas, the statutory Gas Consumers Council.

The Role of the DGs in the Structure and Ownership of the Utilities

Since privatisation, major changes in the ownership and structure of the utility industries have taken place, notably through licensing of new suppliers and mergers and takeovers. The industry DGs have been central participants in decision making over such changes.

The DGs have been in the forefront of pressure to extend competition. In telecommunications, Oftel was closely involved in the government's 1990 Duopoly Review,[58] that extended competition to voice telephony and the infrastructure in the United Kingdom; it then pressed for the inclusion of international telecommunications.[59] In gas and electricity, Spottiswoode and Littlechild strongly advocated widening competition to domestic users, despite opposition from the incumbent monopolists, and also questions from consumer groups as to the benefits for domestic users.[60] Moreover, the two DGs have played a central part in the preparations for competition, such as establishing trials and appropriate computer systems. They have been crucial in decisions on the timing of liberalisation: both DGs have pressed for a rapid transition to competition in the domestic markets, in the face of calls for a slower development and scepticism as to the state of preparations not only by the incumbents, but also by MPs.[61] The importance of the DGs in the extension of competition can be seen by contrasting telecommunications and energy with developments in water: concern by Ian Byatt that the conditions for 'inset appointments' (whereby a water company can establish supply within the area of another water company) were inappropriate meant that in the first six years after privatisation, only one inset appointment was made.[62]

When competition has been extended, the DGs have influenced the terms of new licences: their decisions and advice on matters such as interconnection and universal service obligations have been incorporated into licence terms. When only a few licences were available, the DGs have been important in determining that number and its distribution, especially in telecommunications; thus, for instance, Oftel advised the DTI as to the number of licences for 'Telepoint' services in 1988 and for PCN (Personal Communications Networks) services in 1989 and as to which of the applicants should be selected.[63] Oftel provided a clear framework for choice of licensees (for instance, using PCN to develop serious rivals to BT, and hence not giving Cellnet, in which BT held a 60 per cent stake, a PCN licence) and in being an 'apolitical' expert agency, a method of distancing the government from choosing among large, powerful applicants.

Takeovers and mergers have provided another route for changing the nature of supply in the utilities, especially in electricity and water.[64] The influence of the DGs over change has varied. They have offered advice to the Secretary of State for Trade and Industry over whether he/she should refer bids to the MMC.[65] Sometimes, such advice has not been followed. Professor Stephen Littlechild has had greatest experience of this, in both directions: thus, for instance, his view that Scottish Power's bid for Manweb should be referred to the MMC in 1995 was not followed by Ian Lang, whilst his advice in 1997 that PacifiCorp's takeover of Energy Group did not merit an MMC referral was not followed by Margaret Beckett.[66] On several occasions, however, the DGs have played influential roles both concerning the terms on which a takeover bid has proceeded and in whether it was referred to the MMC. Thus, for instance, DGs have 'reached understandings' with bidders on future price cuts that have allowed the DG to recommend that the bid be allowed to proceed.[67] The secretary of state at the DTI has followed the DGs' views on several occasions in deciding whether or not to refer bids to the MMC – thus, for instance, Littlechild's general argument that ownership by foreign utilities (especially US ones) was not in itself a problem aided the non-referral of several of the bids for regional electricity companies.[68] Finally, when bids have been permitted, the DGs have been important in the conditions imposed on bidders, notably ring-fencing of regulated businesses, publication of separate regulatory accounts for such businesses and provisions to prevent abuse of market position.[69]

The role of the DGs in takeovers has been justified in three ways. First, the DGs may lose information that may be vital for effective regulation. One example is that if separate British suppliers merge, the industry DGs have fewer 'comparators' and hence comparing the performance (and especially, efficiency) of differing suppliers is more difficult; this is particularly important in water, where direct competition between suppliers is difficult, but also applies to electricity. Another example is that if regulated businesses form part of a wider set of activities within a company, it is difficult to discern exploitation of market or anti-competitive behaviour, such as cross-subsidisation between regulated and unregulated activities. Second, mergers and takeovers may lead to suppliers having market power, and damage or prevent fair and effective competition; this argument is strongest in the case of takeovers leading to vertical integration (for instance, between electricity generators and regional electricity companies). Third, however, mergers and takeovers may lead to lower costs and hence the possibility of lower prices for consumers, through the benefits of economies of scale or scope and/or improved management. All three arguments have been used by the DGs, but since they can apply in opposing directions, and since the impact of each can be varied by regulatory measures, they have also permitted considerable discretion for the DGs in the appreciation of the application of the factors to specific situations.

Relations between the DGs and Licensees

The DGs' relations with licensees have been dominated by regulation of the privatised former monopoly incumbents. Since privatisation, relations between the latter and the DGs have been intense, being marked by frequent public disagreements and conflicts. The DGs' activities have become tentacular in terms of the range of matters and the degree of detail, driven by the aims of ensuring 'fair and effective competition' and protecting the consumer, together with the nature of price controls.

Price Controls on Services Sold to Final Users.[70] Price controls applying to services supplied by the privatised suppliers have formed the public centrepiece of the DGs' work.[71] Controls have applied to many of the services offered by the privatised utilities, especially those that are heavily used by domestic customers.[72] Hence, they have affected services worth billions of pounds, as well as influencing the development of competition (future or present). The justification for price controls has been either the existence of legal monopolies or the continuing market power and dominance of the former incumbents.

The form taken by price controls has been licence conditions, set for a number of years. The DGs have been active in revising controls, either at the expiry of the fixed time period, or even during the period (the best-known example being Littlechild's decision in 1995 to re-open the price caps on regional electricity companies that should have applied for the period 1994–98). Modifications of price controls have involved elaborate processes, with consultation documents, proposals and then long contacts between the industry DGs and suppliers, which have constituted a form of negotiation. The major weapon used by the industry DGs has been referral of licence conditions to the MMC, with the suppliers facing the dangers of prolonged uncertainty, a harsher result than that obtainable by agreement with the DG and, most important of all, being broken up. Nearly all modifications of price controls have been made by agreement, with only a handful leading to referrals to the MMC; but, when referrals have been made, the MMC has generally supported the proposals of the industry DGs.[73]

Price controls have often been very detailed: not only have 'baskets' of services been covered, but within these, caps have been placed on prices for individual services.[74] Moreover, the DGs have given 'guidance' as to the evolution of prices within baskets – for example, in the mid-late 1980s,

Oftel indicated that it supported price rebalancing by BT, whereby prices for national and international calls would fall but those for rentals and local calls would rise, but stated that such rebalancing should be gradual. In the late 1980s and early 1990s, the scope of controls was widened;[75] this trend may have been reversed in the mid-1990s, with Oftel claiming that its 1997–2001 price controls are the last on BT's retail prices, and its decision to remove restrictions on BT's line rental charges. Nevertheless, the DGs retain great power over prices, including indirectly through regulation of charges for use of network infrastructures and wholesale supply (see below).

Beginning with controls over BT's prices in 1984, the form of price controls used has been 'RPI-X' (Retail Price Index minus a particular fixed percentage) set for several years. The original aims were to avoid the complexities, and informational and incentive difficulties of price caps based on rates of return, especially given the American experience.[76] Moreover, the RPI-X approach was expected to prevent constant 'interference' by the regulators and to limit their discretion. In practice, however, RPI-X controls have been far from simple and have provided many opportunities for activity and discretion by the DGs. First, when RPI-X periods have run out, the DGs have looked closely at the rates of return earned by the licensee. They have determined what they regard as suitable returns, and also what efficiency improvements they believe are possible, and then sought to set price caps accordingly. Second, when rates of return have been 'excessive' during a particular RPI-X period, the DGs have threatened to modify the price cap, re-opening the formula, and have sometimes done so.[77] Third, other factors apart from the RPI and 'X' have been included in price control formulae, formally, or sometimes informally. Thus, for instance, 'cost pass throughs' have been permitted.[78] In water, after discussions with the water suppliers, Ofwat has set them 'informal' investment targets that were used in calculations for price controls.

The operation of the RPI-X system has required (or perhaps justified) the DGs obtaining information concerning many aspects of the operation of licensees' businesses and taking positions on their measurement and significance. Such matters have included: possible future efficiency gains, rates of return in industries with comparable levels of risk, the cost of capital, investment levels and quality of service.

Access/Interconnection. In order to pursue 'fair and effective' competition, the DGs have regulated access to the utility infrastructures or networks – in particular, BT's network, British Gas' gas pipelines and, in electricity, both the national grid and the RECs' distribution networks.[79] Regulation has been

essential to the development of competition since after privatisation, the
former incumbents have continued to enjoy monopolies or dominant
positions in the supply of the infrastructures on which utility services must
be provided or pass.[80] Whilst the right of third parties to access to the former
incumbents' networks and of interconnection of their own networks have
been licence conditions, translating those rights into practice has been
complex. The DGs have acted on the scope of interconnection,[81] prices (both
overall levels, based on rates on return,[82] and structure, largely based on
costs[83]), non-discrimination (between third parties and the other parts of the
suppliers, which have often been competing 'downstream' with the third
parties) and transfer of information between different parts of the suppliers'
business (since infrastructure suppliers will know the identity of the third
parties' customers).

The Structure and Market Shares of Licensees. 'Conduct regulation' of large
firms with market power has been severely criticised by economists and
contrasted with 'structural regulation' whereby suppliers are broken up into
appropriate units that have limited market power.[84] Yet, the DGs have made
little attempt to break up the former incumbents either through a wide-
ranging reference of the businesses and licence conditions of the suppliers
to the MMC or use of general competition law powers held concurrently
with the DGFT.[85] Instead, they have employed two other strategies. First,
they have sought internal separation of differing parts of licensees'
businesses; thus, for instance, they have required separation of accounts,
staff, management and information.[86] Second, they have taken direct
measures to reduce the market shares of the previous incumbents; thus, for
instance, in the electricity generation market, National Power and
PowerGen, under the threat of a reference to the MMC, agreed with the
DGES to use their 'best endeavours' to sell certain amounts of generating
capacity by 1995, while after the 1991 review by the OFT, British Gas
agreed with the DGGS to reduce its share of the traditional non-tariff gas
market to 40 per cent by 1995.

Preventing anti-Competitive Practices. Action concerning anti-competitive
practices has mostly been taken by the industry DGs rather than the DGFT.[87]
Moreover, they have mostly acted using licence conditions rather than
general competition law. Two sets of measures have been taken. One is a
series of actions against specific anti-competitive practices that have
occurred. Thus, for instance, Oftel has obliged BT to end cross-subsidies
between businesses,[88] to provide access on equal terms to its network and
capacity,[89] and not to discriminate against classes of user.[90] The other is a

move to broader licence conditions that prohibit anti-competitive behaviour. This has been seen in telecommunications, where Oftel imposed a licence change on BT in 1996 prohibiting anti-competitive agreements and abuse of a dominant position, modelled on articles 85 and 86 of the Treaty of Rome, but without the same range of sanctions; indeed, Oftel has declared that it views itself as a 'competition authority'.[91]

Quality and Complaints. The DGs have sought to ensure appropriate quality of service by suppliers, linking it to price controls. They have not only used their powers under the Competition and Service (Utilities) Act 1992, but also informal methods; thus, for instance, Oftel established a 'voluntary' agreement with BT on quality of service; in water, Yorkshire Water agreed with Ofwat in 1996 to forego price rises because of poor service,[92] while Ian Byatt publicly criticised named water suppliers for poor service and has agreed targets for matters such as leakages and discharges with several water companies, with the threat of stricter price controls at the next review if service targets are not met. The DGs have collected and published data on complaints and the quality of services.

Investment. Despite privatisation, investment decisions by suppliers have become an important concern in regulation. The DGs have looked at investment to ensure that their industries develop; suppliers, especially incumbent monopolists, have argued that proper rates of return are needed to allow sufficient investment. Investment has become linked to price controls: rates of return are related to valuations of capital, estimates of future investment and the cost of capital. In water, the DGWS has been directly involved in investment plans: suppliers have agreed investment targets, and when those targets have not been met, Ian Byatt not only publicly criticised companies but also in 1997 agreed with all water companies bar one that they should 'voluntarily' limit their price rises below the maximum permitted and, furthermore, decided to introduce new controls from 2000 to take account of the investment shortfalls.[93]

Costs and Dividends. The DGs have sought to influence both the costs and dividend payments of licensees. Their views, and sometimes approval, have been sought by suppliers over purchases, especially of raw materials and equipment. The DGs have supported decisions that reduce costs, even those threatening employment in Britain, on the grounds of benefits for the consumer and their duty to promote efficiency and economy. Thus, for example, Oftel supported BT's decision in 1985 to buy digital switches from Ericsson,[94] and Offer did not seek to block the move by generators from

British coal to foreign coal and gas-fired generation; in gas, Ofgas put pressure on British Gas publicly to deal with its long-term 'take or pay' contracts for the purchase of gas. Company dividends have not escaped the attention of the regulators: Ian Byatt has publicly attacked 'unsustainable dividends',[95] whilst 'excessive' dividend and other payments to shareholders by regional electricity companies in 1995 (in attempts to ward off potential predators) convinced Stephen Littlechild to re-open price controls.[96]

The Interests of Consumers – Social and 'Public Interest' Objectives. The DGs have argued that extending 'fair and effective' competition serves the 'public interest' and the interests of consumers.[97] As a result, their activities to alleviate the effects of competition have been limited by their conception of the role of competition. 'Social policy' aims have not been given prominence, other than through the benefits of competition for the consumers and measures that involve cross-subsidies between services and/or users have not enjoyed favour. Instead, other methods of dealing with unprofitable objectives have been preferred. Thus, for instance, 'universal service' has been ensured through licence obligations to supply. In telecommunications, where BT bears the obligation, the Access Deficit Scheme, whereby rival operators were to pay BT for its losses on maintaining access and the local network was soon abandoned, and Oftel has questioned whether the cost of universal service is significant for BT.[98] The DGs have sought to protect vulnerable groups: they have participated in persuading suppliers to introduce 'light user schemes', schemes to help the disabled and codes of conduct on disconnections. But, they have accepted the principles of cost-based pricing, even where this means costs rising for certain categories of user (especially as cross-subsidies are reduced) and of non-payers being disconnected, even for fuel and water supplies.

The principle of competition has also guided the DGs' pursuit of wider 'public interest' goals. In communications, Oftel has supported the ban on BT (and other national public telecommunications operators) being able to simultaneously broadcast entertainment services on their telecommunications networks despite calls for the ban to be lifted to encourage the development of a widespread broadband network, and it has consistently opposed any national plan for such a network on competition grounds, such as a national optical fibre network or an 'information superhighway' for schools and colleges agreed between Labour and BT. In energy, levies on suppliers were used to fund energy conservation, notably through the Energy Trust;[99] but, in gas, Clare Spottiswoode largely ended such measures, arguing that they constituted a form of tax that should be determined by parliament, not the DGs.

CONCLUSION

Comparison of regulation of the utilities before and after institutional reform indicates the degree of change that has taken place. The aims and focus of policy have altered, with the replacement of monopoly supply by the pursuit of 'fair and effective' competition. Ministerial action to use the nationalised industries for wider macro-economic and social aims have given way to a raft of measures to pursue 'fair and effective' competition. Cross-subsidisation between services and groups of users has been discouraged and reduced. Change has applied to the processes of policy making: the closed regulatory game of the pre-privatisation era has given way to a more open and public one, with more participants, a higher degree of formalisation of decision-making processes, greater public availability of information, more open conflict and complex manoeuvres involving ministers, the DGs, former monopolists, new entrants, consumer bodies and the MMC.

How and why have these changes taken place? Whilst many reasons could be given – new ideas about regulation and natural monopolies, policy learning (notably from the United States), the changing technological and economic characteristics of the utilities – the present contribution has underlined the role of institutional reform. Privatisation of suppliers played a major role; in particular, its form meant that private monopolies largely replaced public ones, and the resulting dissatisfaction and policy problems offered a pressure for reform of regulation. Nevertheless, privatisation did not provide a new regulatory approach. The legislative framework set out a new division of regulatory functions and powers, and established a set of objectives. But, the legislation was often broad, offered considerable scope for interpretation and did not provide a detailed blueprint for regulation. A crucial, third institutional factor was the establishment of new economic regulators.

The DGs have been central to the changing nature of regulation in Britain in the 1980s and 1990s. From their broad, long and potentially contradictory set of statutory aims/duties, they have fashioned a conceptual framework. They have used that framework both to justify action and to establish clearer, more achievable and coherent goals than had existed in the period of public ownership. Helped by their conceptual framework, the DGs have deployed their statutory powers to the full, taking action across many fields, from price controls and interconnection to quality of service and investment. They have developed a broad interpretation of their role, acting in many domains that at first sight lie outside their intended field of action. Applying their powers and considerable expertise and information, the DGs have altered both the processes of policy making and the forms and extent of public control over suppliers.

Given the analysis of the role of the industry DGs in the utilities, three broader themes can be considered. The first is the extent to which conclusions from the utilities apply to other areas of regulation in Britain. Without attempting a full survey, it does appear that new regulatory bodies have altered regulation and policy making and that some of the features seen in the utilities can be 'read across' to other policy fields. Thus, for instance, bodies such as Office for Standards in Education (Ofsted) or the SIB (Securities and Investment Board) have developed their own conceptual frameworks, often at odds with the traditional philosophy of the sector, engaged in sharp conflicts with powerful established bodies in the area, intervened in great detail over the activities of suppliers, have collected and published new types of data and have instituted decision-making processes that involve consultation and involvement of user groups. While claims of a 'regulatory state' in Britain may be exaggerated,[100] it can be said that the new regulatory bodies have had an impact on regulation and policy making.

The implications of that impact for views of policy making and the British state offers a second broader theme for discussion. Privatisation, the establishment of a new regulatory regime and the creation of the DGs challenge the picture of a Britain stricken by 'institutional inertia'. Regulation of the privatised utilities offers a departure from the style of 'bureaucratic accommodation' and closed policy communities consisting of civil servants and suppliers. It represents an example of policy makers drawing in new participants, taking decisions more publicly and engaging in more open conflict with traditional established interests, phenomena that can be seen as hallmarks of 'Thatcherite' policy making.[101] But, breaking up the closed policy communities has not weakened the state; on the contrary, traditional views of Britain as a 'weak state' are confronted with powerful regulators playing central roles in the supply of services and in the decisions of large companies and able to impose their will, even in the face of opposition of former monopoly suppliers and against traditional patterns of behaviour. The 1980s and 1990s have not seen 'de-regulation' or a 'rolling back of the state'; on the contrary, to the privatised suppliers, the DGs often represent the ever-present and over-powerful state!

Analysis of the role of the DGs aids an explanation of the twin paradoxes of the development of powerful regulators by a Conservative government claiming to 'de-regulate' and the result of powers being transferred from ministers to new bodies being the creation of a more fragmented, but also more powerful state. The industry DGs were established to counterbalance the lack of structural reform at the time of privatisation and to offer reassurance for consumers, but their future role was not foreseen in detail and their creation appeared to be largely a sop to critics of 'privatisations without competition'. But, deploying their powers to the full and developing a conceptual framework and considerable expertise, the DGs have in fact altered the nature of policy making and increased the power of public officials over suppliers. The establishment of specialised organisations, regulating privatised utilities and facing pressures to justify their existence and develop a coherent role, has led to a more powerful state than the previous arrangements under nationalisation.

The impact of the new regulators in the utilities allows consideration of the third theme, the theoretical paradox of institutions and change. Privatisation and the accompanying new regulatory framework showed that institutional reform is possible, albeit only in the presence of the right conditions to provide impetus needed to overcome tradition and indeed opposition. However, the issue then arises of how and why institutional reform led to policy change. To these questions, the British experience of utility regulation provides responses that constitute valuable examples, especially in strategic economic sectors and in a country said to be chained to its traditions. Influential and well-established suppliers were removed from the public sector. New organisations were established, notably the industry DGs. They were generally headed by a fresh group of policy makers who were neither party politicians nor long-serving civil servants.[102] The new regulators broke with previous patterns of policy making, developing their own approach. They sought to define and develop their role and position in regulation. The institutional framework enabled them to do so. It provided them with a set of powers that could be deployed. At the same time, it was flexible, offering scope for interpretation and adaptation. Thus the new institutional arrangements allowed and indeed encouraged policy change.

The theoretical paradox of institutions that aid rather than constrain change is linked to the empirical paradoxes of the changing nature of the British state. Many factors play a role in policy innovation – new ideas and paradigms, policy learning, technological and economic developments, the climate of opinion and the personalities of individual participants. But, such factors pass through the institutional framework of policy making. The British experience of utility regulation suggests that institutions can facilitate change and lead it in particular directions, often unforeseen ones. Institutional reform can, therefore, modify a state's capacities and role in policy making, resulting in a break with the past.

NOTES

The author expresses his thanks to Hugh Berrington, Colin Scott and Vincent Wright for their comments and suggestions on an earlier draft; he also gratefully acknowledges ESRC Fellowship no. H53627502595.

1. For surveys of 'new institutionalism', see P.A. Hall and R. Taylor, 'Political Science and the Three New Institutionalisms', *Political Studies* 44 (1996) pp.936–57; T.A. Koebel, 'The New Institutionalism in Political Science and Sociology', *Comparative Politics* 27/2 (1995) pp.231–43.
2. Cf. S.D. Krasner, 'Approaches to the State: Alternative Conceptions and Historical Dynamics', *Comparative Politics* 21/1 (1984) pp.66–94.
3. K.A. Thelen and S. Steinmo, 'Historical Institutionalism in Comparative Politics', in idem and F. Longstreth (eds.) *Structuring Politics: Historical Institutionalism in Comparative Perspective* (Cambridge: CUP 1992); P.A. Hall, *Governing the Economy* (Cambridge: Polity Press 1986).
4. See, for example, Hall, ibid., E.M. Immergut, *Health Politics: Interests and Institutions in Western Europe* (Cambridge: CUP 1992), F. Dobbin, *Forging Industrial Policy: The United States, Britain, and France in the Railway Age* (Princeton UP 1994).
5. J.E.S. Hayward, 'Institutional Inertia and Political Impetus in France and Britain', *European Journal of Political Research* 4 (1976) pp.341–59.
6. K. Dyson, *The State Tradition in Western Europe* (Oxford: Martin Roberston 1980), Hall (note 3).
7. P. Katzenstein (ed.), *Between Power and Plenty* (Madison: U. of Wisconsin Press 1978); J. Zysman, *Governments, Markets and Growth* (Ithaca, NY: Cornell UP 1983); for sectoral-level analysis, see M.M. Atkinson and W.D. Coleman, 'Strong States and Weak States: Sectoral Policy Neworks in Advanced Capitalist Economies', *British Journal of Political Science* 19 (1989) pp.47–69.
8. Dyson (note 6).
9. Hall (note 3); K. Dyson, 'The Cultural, Ideological and Structural Context', in K. Dyson and S. Wilks (eds.) *Industrial Crisis* (Oxford: Martin Robertson 1983); A. Shonfield, *Modern Capitalism* (Oxford: OUP 1969).
10. G. Jordan and J.J. Richardson, 'The British Policy Style of the Logic of Negotiation?' in J.J. Richardson (ed.) *Policy Styles in Western Europe* (Hemel Hempstead: Allen & Unwin 1982); J.J. Richardson and G. Jordan, *Governing Under Pressure: The Policy Process in a Post-Parliamentary Democracy* (Oxford: Martin Robertson 1979); idem, *British Politics and the Policy Process* (London: Unwin Hyman 1987).
11. For a brief history, see A.I. Ogus, *Regulation. Legal Form and Economic Theory* (Oxford: OUP 1994) pp.6–10; T. Prosser, *Law and the Regulators* (Oxford: Clarendon 1997) pp.32–45.
12. Cf. Oftsed, National Curriculum Council, the Higher Education Funding Council.
13. Cf. the Securities and Investment Board, the Investment Management Regulatory Organisation, the Securities and Futures Authority, and the Personal Investment Authority – see M. Moran, 'The State and the Financial Services Revolution: A Comparative Analysis', *West European Politics* 17/3 (July 1994) pp.158–77.
14. J.J. Richardson, 'Doing Less by Doing More: British Government 1979–1993', *West European Politics* 17/3 (July 1994) pp.178–97; Cf. G Majone, 'The Rise of the Regulatory State in Europe', ibid. pp.77–101.
15. For general discussions mainly in a European context, see G. Majone (ed.) *Deregulation or Reregulation* (London: Pinter 1990); G. Majone, *Deregulating Europe* (London: Routledge 1996); and special issue of *Journal of European Public Policy* 3/4 (1996).
16. The market value of the privatised companies in the four sectors was £89.1 billion on 1 Sept. 1994, representing approximately 12 per cent of the *Financial Times* index – D. Helm, 'British Utility Regulation: Theory, Practice and Reform', in *Oxford Review of*

Economic Policy 10/3 (1994) pp.17–39.

17 For detailed histories, see: L. Hannah, *Electricity Before Nationalisation* (London: Macmillan 1979); D. Kinnersley, *Troubled Water* (London: Hilary Shipman 1988); D. Parker and E. Penning-Rowsell, *Water Planning in Britain* (London: Allen & Unwin 1980).

18. The Gas Act 1948 established 12 area boards and a national Gas Council, but these were amalgamated into British Gas under the Gas Act 1972.

19. Replacing the Central Electricity Authority and regional area boards.

20. In Scotland, the water industry was owned by local government authorities.

21. The Post Office was a government department until 1969, when it became a public corporation.

22. The main statutes were: the British Telecommunications Act 1981; the Post Office Act 1969; the Gas Acts of 1948 and 1972; the Electricity Acts of 1947 and 1957; The Ministry of Fuel and Power Act 1945; the Water Acts of 1973 and 1983.

23. In particular, the minister of the sponsoring ministry – often the Secretary of State for Industry/Trade and Industry, for Energy/Fuel and Power, and in water, also the Secretary of State for the Environment; in Scotland and Wales, the relevant secretaries of state also had powers.

24. Although sometimes the minister was empowered to license other suppliers, as in telecommunications.

25. See notably, T. Prosser, *Nationalised Industries and Public Control* (Oxford: Blackwell 1986); R. Pryke, *The Nationalized Industries: Policies and Performance since 1968* (Oxford: Martin Robertson 1981); W. Robson, *Nationalised Industry and Public Ownership* (London: Allen & Unwin 1960).

26. Amongst the burgeoning general literature on privatisation, see especially, C.D. Foster, *Privatization, Public Ownership and the Regulation of Natural Monopoly* (Oxford: Blackwell 1992) and C. Graham and T. Prosser, *Privatising Public Enterprises* (Oxford: OUP 1991); for accounts of the privatisation of BT, see J. Moon, J.J. Richardson and P. Smart, 'The Privatisation of British Telecom: A Case Study of the Extended Process of Legislation', *European Journal of Political Research* 14 (1986) pp.339–55; K. Newman, *The Selling of BT* (London: Holt Rinehart & Winston 1986); A. Cawson, P. Holmes, D. Webber, K. Morgan and A. Stevens, *Hostile Brothers* (Oxford: Clarendon Press 1990) pp.87–97; for water, see J.J. Richardson, W. Maloney and W. Rüdig, 'The Dynamics of Policy Change: Lobbying and Water Privatization', *Public Administration* 70/2 (1992) pp.157–75; for electricity, see S. Thomas, 'The Privatization of the Electricity Supply Industry', in J. Surrey (ed.) *The British Electricity Experiment* (London: Earthscan 1996).

27. The dates on which at least a majority of shares were sold were: BT – 1984; British Gas – 1986; Twelve new regional electricity companies (RECs) – 1990; National Power and PowerGen – 1991; Northern Ireland Electricity – 1993; the water and sewerage companies – 1989; British Energy (previously, Nuclear Electric and Scottish Nuclear) – 1996.

28. The terms vary between the sectors; 'licence' applies to telecommunications, electricity, and gas after 1995; the term in water is 'appointment', while it was 'authorisation' in gas before 1995; for simplicity, 'licence' is used here for all four industries.

29. In water, the National Rivers Authority was established in 1989, with regulatory responsibilities for the environment and control of pollution of rivers and river areas; following the Environment Act 1995, the Environment Agency was created in 1996, incorporating the NRA, Her Majesty's Inspectorate of Pollution and the local authority Waste regulation Authorities.

30. See S. Littlechild, *Regulation of British Telecommunications Profitability* (London: Department of Industry 1984); see also the parliamentary debates on the Telecommunications Bill, and in particular, talk of ' light rein' regulatory system – speech by Patrick Jenkin, Parliamentary Debates, House of Commons, 1982–83, Vol.33, Cols.38–9.

31. Legislation mostly refers to 'the' secretary of state; the relevant minister is generally the Secretary of State for Trade and Industry/President of the Board of Trade, although other

ministers, notably those for the environment (especially in water), agriculture, Wales, Scotland and Northern Ireland, do have some powers. In the case of appointments the DGT, the DGES and the DGGS are appointed by the Secretary of State for Trade and Industry/President of the Board of Trade, the third thanks to the office holder also currently being Secretary of State for Energy; the DGWS is jointly appointed by the Secretary of State for Trade and Industry/President of the Board of Trade and the Secretary of State for the Environment.

32. For water, that functions are 'properly carried out'.
33. In water, the wording is stronger, being to ensure that suppliers are 'able to (in particular by securing reasonable returns on their capital) to finance the proper carrying out of their functions'.
34. Although the DGs are limited by the common law requirements of judicial review.
35. For further details, see Hansard/EPF, *The Report of the Commission on the Regulation of the Privatised Utilities* (London: Hansard Society 1996) pp.78–9.
36. More formally, if it finds that the matters specified operate or may be expected to operate against the public interest, and specifies adverse effects, and then concludes that licence changes could remedy or prevent those matters.
37. In electricity, the secretary of state can direct the DGES not to proceed with a reference to the Monopolies and Mergers Commission (MMC); in telecommunications, he/she can both prevent a reference, or, if reference is made and a favourable MMC report is produced, can stop the DG from making licence changes, but only on grounds of national security or external relations.
38. For a discussion, see C. Scott, 'Regulatory Discretion in Licence Modifications: The Scottish Power Case', *Public Law* 1997, forthcoming.
39. For fuller details of the enforcement of competition law in the utilities, see Hansard/EPF (note 35) pp.81–82.
40. The major statutory provision specific to the utilities is that mergers between water suppliers with assets over a certain threshold must be referred by the secretary of state to the MMC.
41. Cf. M. Loughlin and C. Scott, 'The Regulatory State', in P. Dunleavy, I. Holliday and G.Peele (eds.) *Developments in British Politics 5* (forthcoming, 1997) p.204.
42. For statements on the benefits of competition for the consumer, see the annual reports of the DGs; for a particularly clear statement about competition, and its relationship to regulation, see the 1985 Report of the Director General for Telecommunications (London: HMSO 1986) esp. paras.1.2–1.5.
43. B. Carsberg, 'Injecting Competition into Telecommunications', in C. Veljanovski (ed.) *Privatisation and Competition* (London: IEA 1989) and B. Carsberg, 'Office of Telecommunications: Competition and the Duopoly Review', in C. Veljanovski (ed.) *Regulators and the Market* (London: IEA 1991).
44. Examples include: the RPI-X formula adjusted only after a period of years, so that suppliers can earn super-normal profits, but only for a limited period, as in a competitive market, where entry would remove such profits; yardstick regulation, especially in water, whereby the DGWS can compare the different levels of efficiency of suppliers in order to impose controls based on the most efficient, mimicking competition driving out less efficient operators.
45. The individual office holders have been: DGT: Sir Bryan Carsberg 1984–92; Bill Wrigglesworth 1992–93 (acting); Don Cruickshank, 1993– ; DGGS: Sir James McKinnon 1986–93; Clare Spottiswoode 1993– ; DGES: Professor Stephen Littlechild 1989– ; DGWS: Ian Byatt 1989– .
46. *Independent*, 12 Feb. 1997 and 26 May 1997.
47. Thus, for instance, Sir Bryan Carsberg, DGT 1984–92 was a qualified accountant and had been involved in establishing the regulatory framework for telecommunications before 1984; Professor Stephen Littlechild was an expert in utility regulation and Ian Byatt was an economist and former senior Treasury official.

48. *Independent*, 5 June 1997.
49. Ibid. 26 May 1997.
50. I. Byatt, 'Water: The Periodic Review Process', in M.E. Beesley (ed.) *Utility Regulation: Challenge and Response* (London: IEA 1995).
51. *Independent*, 3 July 1997.
52. Ibid. 3 April 1997.
53. Cf. Public Accounts Committee, *The Work of the Directors General of Telecommunications, Gas Supply, Water Services and Electricity Supply HC 89* (London: HMSO 1997) and recent reports of the Trade and Industry Committee, notably *Telecommunications Regulation HC 254* (London: HMSO 1997), *Energy Regulation HC 50–1* (London: HMSO 1997) and *Liberalisation of the Electricity Market HC 279–1* (London: HMSO 1997).
54. For instance, via the all-party water committee or PITCOM (the Parliamentary Information Technology Committee).
55. For discussions of the accountability and legitimacy of regulatory regime, see Hansard/EPF (note 35); D. Corry, D. Souter and M. Waterson, *Regulating Our Utilities* (London: IPPR, 1994); Helm (note 16); C. Veljanovski, *The Future of Industry Regulation in the UK* (London: EPF 1993); P. Hain, 'Regulating for the Common Good', in Helm (note 16).
56. For instance, on proposals for BT's price controls and 'fair trading conditions' in BT's licence.
57. Notably the four country Advisory Committees and Committees for the disabled and for small businesses.
58. Department of Trade and Industry, *Competition and Choice: Telecommunications Policy for the 1990s*, Cmnd 1461 (London: HMSO 1991).
59. *Financial Times*, 20 Nov. 1995; the BT–Mercury duopoly in international communications was ended in 1996.
60. See, for instance, National Consumer Council, *Paying the Price* (London: HMSO 1993) esp. pp.120–32 and National Consumer Council, *Competition in the Electricity Market* (London: NCC 1996).
61. See, for instance, Trade and Industry Committee, *Aspects of the Electricity Supply Industry, HC 481-I* (London: HMSO 1995), *The Domestic Gas Market, HC 23* (London: HMSO 1995) and *Liberalisation of the Electricity Market* (note 53) and references therein to evidence submitted by consumer bodies.
62. *Independent*, 21 Feb. 1997 and 29 Jan. 1997.
63. Interviews; *Financial Times*, 4 Jan. 1989 and 12 Dec. 1989.
64. For a discussion of the electricity industry, see M. Parker, 'Competition: The Continuing Issues', pp.228–32 and J. Surrey, 'Unresolved Issues of Economic Regulation', pp.250–52 in Surrey (note 26).
65. It must be remembered that the secretary of state is obliged by statute to refer mergers between water companies to the MMC, subject to an asset test.
66. *Financial Times*, 1 Sept. 1995 and 2 Aug. 1997.
67. For instance, Stephen Littlechild made the sale of power stations by PowerGen a condition for its bid for Eastern (although the bid was then referred to the MMC) in 1996, whilst in 1994, he 'suggested' the idea of customer rebates to make Trafalgar House's bid for Northern Electric more 'acceptable' to him – *Financial Times*, 25 April 1996, 26 April 1996, 21 Dec. 1994, 22 Dec. 1994 and 24 Dec. 1994; in the bid by Lyonnaise des Eaux for Northumbrian Water, Ian Byatt played an important role, first in his submission to the MMC (following the mandatory reference to it) and then after the MMC report, in his advice to the secretary of state, notably concerning the price cuts that could be introduced after 1999 that he had discussed with the Lyonnaise des Eaux – *Financial Times*, 8 Nov. 1995, 28 Oct. 1995 and 29 Sept. 1995.
68. There were no fewer than seven successful bids by American companies of RECs between 1995 and July 1997 – *Independent*, 2 Aug. 1997.
69. Such conditions have applied most clearly in takeovers involving vertical integration

(notably Scottish Power's takeover of Manweb), horizontal integration (particularly North West Water's takeover of Norweb, forming United Utilities) – see *Financial Times*, 1 Sept. 1995 and 12 Sept. 1995 – as well as bids for RECs.

70. Regulation of use of network infrastructures is considered below under access/interconnection.

71. For an economic analysis of price controls, see M. Armstrong, S. Cowan and J. Vickers, *Regulatory Reform: Economic Analysis and British Experience* (Cambridge, MA: MIT Press 1994) Chs.7–10; R. Rees and J. Vickers, 'RPI-X Price-Cap Regulation', in M. Bishop, J. Kay and C. Mayer (eds.) *The Regulatory Challenge* (Oxford U. Press 1995); for a general analysis of price regulation, see C.D. Foster, *Privatization, Public Ownership and the Regulation of Natural Monopoly* (Oxford: Blackwell 1992) pp.186–25 and A.I. Ogus (note 11) pp.305–17.

72. For instance, the 'tariff' or 'franchise' markets in gas and electricity – that is, supply below certain thresholds – and telephone services (as opposed to advanced telecommunications services).

73. The major examples of references to the MMC concerning price controls on services for final users have been: British Gas in 1992, as part of a wider set of references on British Gas; Scottish Hydro-Electric in 1994; South West Water and Portsmouth Water in 1994; MMC reports were published the following year; it should be noted for the water companies, the MMC was making 'price determination inquiries' rather than licence modification ones, since the water licences ('appointments') specify that the DGWS must refer disagreement on the K 'adjustment' factor to the MMC; for discussion, see H. Liesner, 'The Role of the MMC in Utility Regulation' in Helm (note 16). The exception to harmony between the MMC and the DGs has been in Northern Ireland, where the MMC's recommendations on price controls on Northern Ireland Electricity in 1997 have been rejected by Ofreg, the electricity regulator for the Province – *Independent*, 27 April 1997 and *Financial Times*, 7 Aug. 1997.

74. For instance: rentals for telephones for BT; standing charges for British Gas and infrastructure charges for water supply companies – see Rees and Vickers (note 71) pp.374–5.

75. For instance, BT's charges for connection charges were the subject of an 'informal agreement' with Oftel in 1988, whilst international calls were included in the basket in 1991.

76. See Littlechild (note 30).

77. The clearest example was Littlechild reopening the cap for the regional electricity companies in 1995, so that the new caps for 1995–2000 was re-determined in 1996; for BT, the price cap was widened in 1991 to include international calls, thereby altering 'X', which had been fixed in 1989; for British Gas, following the 1992 reference to the MMC and report in 1993, the price cap for 1992–97 was re-determined in 1994; for a summary of changes in RPI-X caps, see House of Commons Trade and Industry Committee, *Energy Regulation*, HC50–1 (London: HMSO 1997), p.xix, and House of Commons Trade and Industry Committee, *Telecommunications Regulation*, HC 254 (London: HMSO 1997).

78. For instance, of gas purchase costs (for British Gas prices for tariff customers), fossil fuel and nuclear levies for the regional electricity companies, and in water, of the costs of unforeseen tightening of environmental and quality standards of estimates of the cost of capital required – see Armstrong (note 71).

79. Since there have been almost no inset appointments in water, access to the mains networks has not been significant.

80. For a discussion of the economic characteristics of networks, see J. Vickers and G. Yarrow, *Privatization: An Economic Analysis* (Cambridge, MA: MIT 1988), and also Armstrong, Cowan and Vickers (note 71).

81. Most notably in telecommunications, where Oftel determined in 1985 that Mercury should enjoy full access to BT's network, even if this meant that BT's local network was used at both ends of the call.

82. The subject of considerable conflict, especially in gas: in 1992 and 1996, British Gas and Ofgas failed to agree on appropriate accounting methods and rates of return, leading to

references to the MMC, which largely supported Ofgas' position – see the two reports of the MMC in 1993, Monopolies and Mergers Commission, *Gas*, Vol.1, Cm. 2314 (London: HMSO 1993) and *Gas and British Gas*, 3 Vols, Cm. 2315, 2316 and 2317 (ibid.) and *British Gas: A Report under the Gas Act 1986 on the Regulation of Prices for Gas Transportation and Storage Services* (London: HMSO 1997).

83. Cf. Oftel, *Determination of Terms and Conditions for the Purposes of an Agreement on the Interconnection of the BT System and the Mercury Communications Ltd System* (London: Oftel 1985).

84. For critiques of conduct regulation, see Vickers (note 8) and Armstrong et al. (note 71).

85. Ofgas made the widest reference to the MMC of British Gas in 1992, but the MMC's recommendation that British Gas be forced to divest itself of trading was rejected by the government.

86. For instance, for British Gas, between TransCo (transportation ad storage businesses) and its trading businesses, and for water companies, between their regulated businesses and other activities.

87. The main exception being OFT referrals of the non-tariff gas market to the MMC in 1987 and its review in 1991.

88. For instance, of its value-added network services and customer premises equipment businesses.

89. For example, for transmission links for satellite services or for voicemail services – Oftel, *Representation on Behalf of PanAmSat* (London: Oftel 1988).

90. To appreciate the breadth and detail of Oftel's work on anti-competitive practices, see its *Competition Bulletin* with details of its case work.

91. Cf. C. Scott, 'Deregulation of BT's Pricing and the New Fair Trading Requirements', *Utilities Law Review* 7 (1996) pp.176–7.

92. *Financial Times*, 4 June 1996.

93. *Independent*, 31 Jan. 1997 and 13 Feb. 1997.

94. Oftel, *British Telecom's Procurement of Digital Exchanges* (London: Oftel 1985).

95. *Independent*, 18 June 1997.

96. For a discussion, see Prosser (note 11) pp.164–5.

97. For clear statements, see example, the 1984 *Report of the Director General of Telecommunications* (London: HMSO 1985) paras. 1.2–1.6; evidence of Clare Spottiswoode and Professor Stephen Littlechild, Trade and Industry Committee, *Energy Regulation* (note 77) para. 32.

98. Oftel, *Universal Telecommunication Services – Proposed Arrangements for Universal Service in the UK from 1997* (London: Oftel 1997).

99. In gas, British Gas was allowed to pass on the costs of investment in energy efficiency measures, known as the 'E factor' – Cf. Trade and Industry Select Committee, *Energy*, (note 77) paras. 115–28.

100. See Loughlin and Scott (note 41).

101. Cf. Richardson (note 14).

102. The exception is Ian Byatt, who was formerly a senior Treasury official; Clare Spottiswoode had also spent a short time in the civil service.

The Judicial Dimension in British Politics

NEVIL JOHNSON

During the past 20 years or more the decisions of judges and of other officeholders and institutions with adjudicatory functions have begun to impinge more often and more insistently on the spheres of political and administrative discretion in Britain than was usual in the past. Moreover, determinations of the courts, and especially in the sphere of public law, receive a degree of public attention virtually unknown a generation ago. It is not uncommon for two or three decisions affecting the powers and financial liabilities of the government or other public authorities to come out within a single week and to be commented upon in approving terms in the media. All this stands in sharp contrast to the traditional understanding of the role of the courts. When the independence of the judicial power was conclusively established in England by the 'Glorious Revolution' of 1688–89, one of the implicit conditions of this outcome was that the judges should take care not to encroach on the legitimate spheres of discretion of the sovereign political authorities, Crown and Parliament. Judges were expected to interpret the law – both the judge-made common law and parliamentary statutes – in accordance with precedents and reason. It was not their job to shape public policy, even though occasionally by their decisions they did establish great points of principle central to the rule of law, as for example in *Entick v. Carrington* in 1765.[1] But for the most part the judges of the common law did not insert themselves into the spheres reserved for the discretion of the Crown and its ministers or of Parliament as the sole source of new statute law.

During the present century both of the main tendencies in British party political life have generally endorsed this habit of judicial reserve. The Right has been content for judges to stick to a literal interpretation of statutes and uphold the strong bias in favour of the rights of private property inherent in the common law.[2] Such an approach also meant that the prerogatives of both Parliament and ministers of the Crown remained as a rule beyond serious challenge in the courts. All this was broadly in harmony with the Conservative preference for strong, though limited, government. But the Left, too, has usually preferred to keep judges and legal processes out of the traditional spheres of politics and government, though for somewhat different reasons from those which carried weight with Conservatives. For the Left, and especially for the Labour Party after it succeeded the Liberals as the principal party of reform, the judicial power was an object of suspicion on several grounds and should be kept at bay. It was plainly 'undemocratic' and on that account alone should not impose its values on decisions reached by democratically accountable politicians; it was persistently prejudiced in favour of private-property rights and could, therefore, be expected to oppose collectivising measures involving wealth redistribution and greater equality; and more generally, since judges and many lawyers came from privileged levels of society, they could not (like politicians) understand the aspirations of 'ordinary people'. So the traditional view from the Left was that the less judicial interference there was with the interpretation of statutes, the better it would be for the cause of social reform.[3]

In recent years, however, these once familiar political attitudes towards the judicial power have been rendered largely obsolete simply as a result of the steady growth in judicial activism, the increasing propensity in society to take issues to a court for resolution, and the willingness of ministers and Parliament to tolerate more checks to their discretion than would have been regarded as acceptable 30 or 40 years ago.[4] Conservative governments after 1979 often enough found their reforming zeal checked by judicial decisions which sought to enforce procedural restraints on executive action, and thus got in the way of changes deemed both necessary and desirable. And on the Left the revived Labour Party under Mr Blair appears positively to welcome judicial interventions and has committed itself to extend the remit of the courts by incorporation of the European Convention on Human Rights and Fundamental Freedoms into British domestic law. Alongside all this there has been the slow but steady widening of the jurisdiction of the European Court of Justice (ECJ), the ultimate arbiter of legal disputes falling within the sphere of the European Community (EC) and European Union (EU).

So we face at the outset something like a double paradox. Conservatives committed to 'rolling back the state' and to increasing the responsibilities of individuals have none the less sometimes found the courts standing in their way; out of power until recently the Labour Party has come to take a much more sympathetic view of what judges do and even looks set to widen their scope for taking decisions likely to cut down further the sphere of ministerial discretion. Such a widespread acknowledgement of the range and penetration of contemporary judicial decision making suggests that the explanation for the change cannot be found simply in judicial reponses to the political constellation of the 1980s. There must be wider-ranging developments which have worked in favour of looking more often to the judicial power for the resolution of arguments and conflicts of interest. It is indeed my contention that this is so, but it will be better to come back to these wider changes after we have looked more closely at some of the specific developments in Britain in recent years which have brought the judges more often into areas of controversy and political argument.

THE EXPANSION OF JUDICIAL REVIEW

The somewhat agitated attention paid to several fairly recent judicial decisions tends to obscure the fact that the beginnings of a more active role by judges in the review of governmental actions lie far back in the 1960s. Cases such as *Ridge v. Baldwin* (1963), *Rookes v. Barnard* (1964), *Padfield v. The Minister of Agriculture, Fisheries and Food* (1968) and *Anisminic Ltd v. Foreign Compensation Commission* (1969) represent but a part of the flow of judicial decisions which reversed the attitudes of extreme deference to executive discretion which had generally been dominant for 30 years or more prior to 1960. Much of the credit for this shift to a more ingenious use of procedural values in order to impose more stringent standards of reasonableness and fairness on public bodies must go to Lord Reid, a leading member of the judicial committee of the House of Lords for a remarkably long period stretching from 1948 to 1974.[5]

Further decisions during the 1970s, notably the *Secretary of State for Education and Science v. Tameside Metropolitan Borough* (1976) and *Laker Airways v. the Department of Trade* (1977) extended still further the judicial claim to examine the use of ministerial discretions and to subject them to qualifying conditions, even when unprescribed by the relevant statutes. Conflicts provoked by the actions and claims of trade unions and their members during the 1970s and later also brought the judges more often into politically contentious matters and this was to prove the harbinger of what looks like an enduring shift towards a pattern of industrial relations more highly regulated and more frequently guided by judicial conclusions than anything contemplated before 1970. Here the move towards a more formalised system of adjudication involving both industrial tribunals and from time to time the courts was in harmony with legislative initiatives taken first by the Heath government in 1971 (subsequently largely repealed by a Labour Party in thrall to the trade union barons), and then more skilfully and effectively by the Thatcher governments of the 1980s.[6]

The other major step in the direction of reinvigorating judicial review was the procedural reforms of the period 1977 to 1981, prompted by the Law Commission, carried forward by the Supreme Court itself, and culminating in the Supreme Court Act 1981 which gave statutory force to some of the changes made. Essentially what these changes did was to simplify access to available legal remedies in public law disputes by providing for a standard form of application for judicial review. Existing remedies by prerogative writ and by a variety of other means were not abolished, but instead the procedure for invoking a legal remedy was simplified and became easier to understand. It is clear that one of the effects of these changes has been a steady rise in the rate of applications for judicial review both on the part of private individuals and of public bodies engaged in disputes with other public bodies. Available judicial statistics are not particularly helpful in facilitating measurement of this increase in judicial review applications. They do, however, indicate quite clearly that from a total of 685 civil and criminal applications in 1982 the number of applications for leave to apply rose to 2,129 in 1990 and to 2,886 in 1993. Corresponding figures for leave granted in the same years were 468, 902, and 1,204.[7] One crucial factor pushing up the rate of application for judicial review has been the rapid development of pressure groups eager to act on behalf of potential litigants and specialising in particular fields of law such as immigration and nationality law, one of the most fertile areas for challenges via applications for judicial review.

It is difficult to trace any clear line of doctrinal development on the part of judges in their approach both to the initial procedural stage of requesting judicial review and then to the substantive issues arising if a case goes forward for judgment. But there is some evidence to suggest that they steadily became more adventurous during the 1980s and were keen to take opportunities to extend further the procedural limitations inherent in the principles of legality, procedural propriety, reasonableness and, most recently, proportionality which may be held to restrict the scope of ministerial discretion. By 1993 the courts were ready to deny in *M v. Home Office* that there was no power to enforce the law by injunction or contempt proceedings against a minister acting in his official capacity,[8] and a couple of years later in the Pergau Dam case the judges came very near to censuring the Foreign Secretary for what was essentially a policy decision.[9]

Judges, just like other groups in society, respond to its political evolution and take account of shifts in public policy. What seems to have happened during the Thatcher era and later is that gradually judges – or at least some of them – fell out of sympathy with the pace, style and at least some of the objectives of the economic reform project to which Mrs Thatcher and her governments were committed after they took power in 1979. Change in the shape of new policies and new legislation was frequent, radical, upsetting to many vested interests and sometimes ill-thought-out. Not surprisingly this prompted many hostile reactions and among these was a readiness to get involved in legal controversy and litigation in the hope of holding up some of the changes being made. Foremost amongst the institutionalised objectors were trade unions and local authorities. The latter in particular were subjected to an unprecedented flood of legislation from 1980 onwards, much of it designed to control their spending, to limit and modify their sources of revenue, to revise their powers or to subject the exercise of these powers to a variety of new and uncomfortable constraints.

This is by no means to suggest that judges rushed like knights in shining armour to the defence of local authorities. Indeed, in *Derbyshire County Council v. Times Newspapers* in 1993 the Court of Appeal ruled that neither a local authority nor a government agency could sue for libel, since to do so would constitute an unnecessary interference with free speech,[10] while ten years earlier Lord Denning had come down heavily against the former Greater London Council in its dispute with Bromley over the 'fares fair' policy.[11] As so often has happened in the past, it was not too difficult for the government to tighten its grip over local spending and other aspects of local government activity simply by making the relevant statutory provisions as 'judge-proof' as was humanly possible. In some spheres, however, there was a better prospect of successfully challenging the use of executive authority. Immigration proved to be a fertile field for legal challenge to Home Office decisions and there has too been a significant rise in the number of occasions on which the police have been exposed to judicial proceedings. Sometimes this has involved allegations of misconduct, sometimes failure to perform effectively their functions in relation to the preparation of cases against offenders. The outcome has been that neither the police authorities nor individual members of local police forces enjoy that degree of protection from serious investigation of their conduct which previously prevailed.[12] This has been particularly important in relation to alleged cases of misconduct by the police affecting members of immigrant communities.

The pursuit of contentious policies by the governments of the 1980s undoubtedly stimulated increased litigiousness on the part both of public bodies and individuals affected by them. There was too an increased awareness in society that it was possible to challenge decisions taken both by governmental organisations and often enough by private bodies too, and to have a prospect of securing some recognition of the claims advanced in such proceedings. All this contributed to the stimulation of a discussion of what were taken by many to be the weaknesses of the British constitution, and in particular of the need for more explicit limits on the discretions enjoyed by the Executive and more comprehensive protection of individual rights. Naturally much of this argument was dominated by a range of pressure groups overtly campaigning for constitutional change.[13] But indirectly at least the judiciary also became involved in this process of questioning traditional methods of handling what used to be called 'the redress of grievances'. Indeed, a few judges made no secret of their belief that the time had come for a more active judicial role both in controlling executive action and in imposing policy limits on what governments (and by implication Parliament) were entitled to do. Sometimes such views found expression in actual legal judgments, more often in articles or speeches. Sir John Laws, Sir Stephen Sedley and Lord Woolf (now Master of the Rolls) have been to the fore in expressing opinions favourable to a reinforcement of judicial powers, though with considerable differences of emphasis among them.[14] But it has to be remembered that such public advocacy of the desirability of a more explicit commitment to the enhancement of individual rights is well within the traditions of the British judiciary: today's radicals are following in the footsteps of Sir Leslie Scarman (as he then was) who in the Hamlyn lectures advocated substantial changes in English legal thinking nearly a quarter of a century ago.[15]

It is worth noting, however, that the judiciary reveals nothing approaching a monolithic view either of its role or of the procedural values which it must apply and develop. That this is so reflects the independence of the British legal professions from which the higher levels of the judiciary are recruited as well as the relative diversity of outlook and experience within them. Unlike the situation in several continental European countries where the judiciary is a professional branch of the state service, strongly influenced by academic lawyers, and often enough subject to the unifying pressures of a constitutional law jurisdiction, there is in Britain virtually no basis for the emergence of anything like a dominant orthodoxy or doctrinal uniformity. Law and legal opinion remain open to development and to the need to respond to changes in public perceptions and public policy.

JUDGES AS INVESTIGATORS

The involvement of judges in matters of political concern and public controversy is often seen as a one-way street – the judges getting involved in politics. But it is important to remember that the traffic runs in both directions in the sense that politicians often turn to judges to help them solve difficult problems with which they are faced. Moreover, this is by no means a new phenomenon. It has for a considerable time been usual to call on the services of a High-Court judge to preside over inquiries into events widely regarded as national disasters, for example the Aberfan tragedy in 1966 or the Hillsborough football stadium collapse in 1989. This category of judicial public service is, however, relatively uncontentious, though sometimes such inquiries do have wide-ranging policy implications which lead to political controversy. But the main concern here is with the habit of calling on judges to look into cases of alleged political misconduct. In this connection too there are precedents from long ago. During the Labour government of 1945–51 judges were occasionally appointed to look into such cases of alleged misconduct in public life, while the Macmillan government also turned to judges to look into matters which were both legally and morally delicate, for example the conduct of John Profumo.[16] Under Mrs Thatcher Lord Justice Scarman was asked in 1981 to inquire into the causes of the Brixton race riots and to make recommendations,[17] and his findings had substantial policy consequences both for the government's approach to the handling of discrimination against ethnic groups and for the police in their law and order role. But on the whole the Thatcher period was marked by some reluctance to appoint external committees of inquiry, since to do so was regarded as often no more than a device for shedding governmental responsibility and buying time.

After John Major took office in 1990 a shift of attitude occurred and this became more marked after 1992 as his government ran into a variety of difficulties resulting in widespread public criticism. The response was an increasing tendency to call on judges to examine what were essentially questions of political conduct and ethics. Perhaps the most striking instance of this was the appointment of Lord Justice Scott, the Vice-Chancellor, to inquire into what became known as the arms for Iraq affair. After an inordinately long and highly public inquiry Sir Richard Scott produced a report which, though exhaustive and of formidable detail, contained little in the shape of definite attributions of culpability.[18] Nevertheless, since the inquiry was a protracted public process and highly damaging for the government, it did tend to confirm in the public mind the belief that ministers had over a considerable period knowingly disguised changes of policy, concealed these from the House of Commons, and come near to allowing innocent businessmen to go to gaol for alleged breaches of export licensing rules. More profound in its longer-term institutional significance was the decision of John Major in October 1994 to establish under Lord Nolan, another senior judge, a committee charged with keeping under investigation and review standards in public life. This step was prompted chiefly by allegations made against various Conservative MPs. The principal immediate effect of the initial Nolan report was the government's decision to accept the recommendation that self-regulation by the House of Commons of the conduct of its members should cease through the appointment of a commissioner for standards and the introduction of more stringent rules governing the declaration of financial interests.

It is understandable that ministers should in morally delicate matters turn to investigators who are held to be objective and impartial rather than attempt to deal with them through conventional political procedures. What is somewhat more surprising is the growing tendency to regard the conclusions of outsiders, and in particular of judges, in such matters as more or less sacrosanct, beyond challenge. This points to one of the dangers of the judiciary's willingness to become involved in inquiries of this kind. This is that they will be used by politicians to legitimate in various ways what the latter then decide to do. Rightly or wrongly judges are held to have the reputation of being immune to political bias and special interest pleadings, and this means that politicians who have got into a hole are easily tempted to conclude that it will help them to get out of this unenviable situation if they can cite the findings of a judicial personage in favour of what they propose to do. Mrs Thatcher was not as a rule inclined to take that way out of a problem, but her successor found it convenient to do so on several occasions. On the other hand, the use of judges to legitimate executive decisions should not be exaggerated. When policy issues of a general nature arise a government may well be ready to override the conclusions of a judge. A clear recent example was provided by the treatment of Lord Cullen's inquiry in 1996 into the massacre by a deranged gunman of school children and their teacher in Dunblane. In this instance the government of the day felt free to go beyond his policy recommendations on the ownership and use of handguns, while its successor has gone even further in bowing to what is held to be the dominant public view on gun control.

It is worth noting that the resort to judges for conclusions free of political bias is far more common in Britain than elsewhere: the professional state judiciaries of continental Europe rarely step outside their prescribed roles, which may, of course, include that of acting as examining magistrates and a prosecuting authority. In the United States, too, it is more likely that officials with experience as state attorneys and prosecuting counsel would undertake inquiries than practising judges. But in Britain judges appear to be all too ready to assume roles which in reality bring them into very close association with the executive power and with issues of often controversial public policy as well as exposing them to the pressures exerted by the public at large.

THE EUROPEAN DIMENSION OF JUDICIALISATION

No reference has so far been made to an important external factor bringing the judicial resolution of conflicts of interest more often into play on the political stage. This is the presence of two European judicial instances to which appeal can (and in some cases must) be made. These are the

European Commission and Court in Strasbourg, established to uphold the European Convention on Human Rights and Fundamental Freedoms to which Britain is a signatory, and the European Court of Justice (ECJ) in Luxembourg which is the judicial arm of the European Community and Union.

The second of these is the guardian of an emerging legal order to which Britain belongs by virtue of the European Communities (EC) Act of 1972. It was clear from the start that entry into the European Communities (as they then were) would have legal consequences, though at the outset it was far from clear how extensive these might be and when they would occur. Even now, 25 years on, the impact of Community law and of decisions made in Luxembourg on domestic law is fairly limited and by no means obvious to most citizens. This is in part because the jurisdiction of the court is limited to matters coming within the scope of the treaties, in part because Britain has a good record in complying with EC legal requirements. But it is also due to the fact that the British courts have been cautious in taking account of the implications of European law.

Legally and constitutionally British acceptance of European law rests on the 1972 legislation bringing Britain into the Communities. It remains the case that Parliament could in principle repeal that legislation or even now legislate to exempt this country from particular legally-valid measures passed within the Community framework, even though other member states might regard this as a breach of treaty obligations. To this extent the principle of parliamentary sovereignty has been preserved, and British courts have not so far ventured to call that conclusion into question. Nevertheless, they have gone some way towards affirming the supremacy of European law to the extent that domestic rules are found to be in conflict with it. This occurred most notably in the *Factortame* cases of 1990 and 1991 in which the courts decided that where there was a conflict between domestic and European legal requirements, it had to be assumed that Parliament wished to give precedence to Community law.[19] Even in this case, however, it appeared to be accepted by the judges that if Parliament were expressly to override European provisions they would have no alternative but to accept that as binding law. While a dramatic head-on collision with the doctrine of parliamentary sovereignty has so far been avoided, there is no doubt that the exposure to European legal concepts, methods of interpretation, and court rulings is having a steady impact on the thinking of British domestic courts. In particular, they are having to face up to the stronger emphasis which European jurisdictions, including the ECJ, place on the formal principle of equality and consistency of treatment. This has already had important consequences for the rights of women in employment, for example, and it is likely that other forms of discrimination will fall foul of the ECJ's reluctance to follow the British preference for treating cases on their merits and thus for tolerating a variety of different outcomes.

In some respects, however, the impact of the European Convention on Human Rights and of the institutions in Strasbourg has been greater, even though the number of cases decided is small and the decisions of the court are not directly applicable in the law of the United Kingdom. But British governments have accepted an obligation in international law to comply with the Convention and they have generally been ready to act on the findings of the Strasbourg court, even though this has sometimes required amending legislation or measures to compel parts of the United Kingdom to fall in line with policies to which they were opposed (notably in the case of the Isle of Man where birching as a punishment had to be given up following an appeal under the Convention). The crucial feature of the European Convention is that it does provide a code of human rights which can be applied in a general way and interpreted according to circumstances. No doubt when Britain signed the Convention (after playing a major part in drafting it) few people thought that it would ever have much application in a country with such a long history of the rule of law and respect for basic freedoms as Britain.[20] But that was to underestimate the dynamic potential of this kind of general code in what has become a highly rights conscious environment. Today few limits can be set to the ingenious application of abstract principles of individual liberty and equal treatment to the extension and justification of individual human rights claims. This was the experience of the Thatcher years and after when more and more litigants found that they might be able to secure in Strasbourg satisfactions denied to them under the domestic jurisdiction. So it was that Britain provided a relatively high proportion of the cases referred to the court in Strasbourg, thus prompting the often unjustified conclusion that British law was peculiarly inadequate in its protection of a range of individual rights.[21]

A more mundane reason for the growing number of pleas to Strasbourg is almost certainly to be found in the more vigorous litigiousness of clients, their lawyers and the pressure groups acting on their behalf, all of whom saw here new possibilities of securing vindication of the claims they put forward. But the fact that this court too, like the ECJ, follows continental European methods of legal interpretation, rather than those familiar in the common-law tradition, has also played a part in producing results often at odds with those which a British court might have favoured. Sometimes, too, the findings of the Strasbourg court have prompted irritation in Whitehall and during the 1980s there were occasionally mutterings about the possibility of modifying or even terminating the right to petition the court. But no such step was taken and the government which came to office in May 1997 is, in contrast, committed to embodying the Convention into British law so that the domestic courts will themselves be mainly responsible for its interpretation. The Human Rights Bill presented to Parliament in late 1997 to give effect to this promise follows in some degree the cautious New Zealand model that allows legislation to prevail over the Bill of Rights.[22]

The overall effect of the possibility of appeal on human rights issues to the Strasbourg court has been to encourage the belief that there is a sympathetic judicial body outside the United Kingdom before which claimants can contest the decisions of both the executive arm of government and of the national courts. Awareness of the fact that this court may well apply general principles capable of broad and flexible interpretation to the cases presented to it, and without much concern for the specific context and circumstances in which the alleged breach of rights has occured, is by now widely diffused, and especially in the legal profession. Furthermore, the respectable record of compliance with Strasbourg findings by British governments further encourages those who hope that they may be successful in prosecuting a claim at this level. It is factors such as these which almost certainly have contributed more to the growing tendency in recent years to 'appeal to Strasbourg' than any political decisions of British governments or systematic deficiencies in the protection afforded to human rights under normal British law.[23] And above all, as will be noted again below, this development finds its place within a movement widespread throughout Western societies to pursue the protection and enhancement of individual and minority rights more or less regardless of the implications this may have for existing legitimate interests as well as for the maintenance of general social stability and harmony.

THE SHIFT FROM DISCRETION TO RULES AND REGULATION

The counterpart of increasing judicial activism in the past 20 years or so is the move away from reliance on political and administrative discretion – usually exercised by ministers and their civil servant agents – to a situation in which the responsibilities for executive action are more widely diffused and, in some fields, made subject to formalised procedures, to the intervention of independent adjudicatory bodies, or to new forms of regulation and supervision such as now apply to the privatised utilities. At first sight these developments do appear paradoxical when set against the spate of policy changes and central government instruction and guidance which took place in the 1980s and after. Here was a government committed to more powerful central political direction and to a range of radical legislative measures affecting many sectors of economic and social policy and service provision. Moreover, the centralising effects of this drive, not least in the areas for which local authorities had traditionally been chiefly responsible, marked a sharp break with the past, for example in the insistence on competitive tendering for many local services and, more significant still, in the measures taken to introduce national curricula and standards in schools.

This paradox is, however, not too difficult to explain. The reforming programme of the Thatcher years certainly depended on a reinforcement of central political direction and called for an extensive programme of legislative change. To that extent there was in comparison with earlier years a strengthening of the central government and a readiness to disturb vested interests not seen for a long time.[24] But the effects in the longer term of many of the changes made were not so much to increase ministerial powers as to change the terms on which other public institutions, officeholders and social organizations had to operate. In the sphere of schools, for example, local authorities saw their own powers reduced in favour of school governors and teachers; in the National Health Service there was, despite the complaints and criticisms of 'more bureaucracy' subsequently made, an enhancement of the responsibilities of hospital managers and of a significant proportion of general practitioners, all of which implied less direction from the centre and from its administrative agents; inside the central administration itself there was put in hand a far-reaching process of delegation of responsibility to executive agencies, a major consequence of which has been that ministers and their immediate administrative staffs are removed from the day-to-day provision of many services to an extent not known 20 years ago.[25]

A general consequence of this loosening up of administrative structures and the introduction of new framework conditions in many fields of provision is that there is more chance of the citizen being able to identify who is responsible for specific decisions than there used to be. This in turn was reinforced by the Citizen's Charter movement formally launched in 1991 as a programme for setting and establishing standards of service which various providers are expected to achieve. Notwithstanding the weaknesses that can be detected in all these initiatives, they have served to underline the fact that services are there for users and 'customers' (as they are so often called now), that they should respond to customers' needs, and that the users are entitled to complain and to seek redress, including compensation, if there are serious shortcomings in the services provided. All this contributes to the more critical environment within which resort to legal remedies has also become more popular and, often enough, more successful than it used to be in the past. What is more, even the justice system itself has not escaped serious criticism and challenge. During recent years several examples of miscarriages of justice have been revealed and acknowledged, resulting not only in the payment of compensation to the victims, but also in a considerable erosion of confidence in both the courts and the police and prosecuting authorities.[26] In parallel, and fed by the media, there has developed a growing tendency for those involved in legal proceedings of various kinds to comment on and often condemn the outcomes, whether these be the verdicts, the levels of punishment set, decisions regarding liability and costs, or simply the procedures as such. Such developments may not enhance respect for the justice system or confidence in it. Yet, they do at the same time indicate a greater readiness on the part of those caught up in legal proceedings to treat the system instrumentally and to try to get as much out of it as possible. It is difficult to avoid the conclusion that the rapid rise in the amount of legal aid granted to litigants in recent years has had some connection with this trend.

Apart from this wider context of judicialised and legally regulated procedures it is worth noting two other indicators of the movement away from informal and discretionary methods of administraton. One is the emergence of numerous regulatory regimes, chiefly for the privatised utilities like British Telecom or the water utilities, but also for quite new agencies such as the National Lottery for which a private-sector company has the franchise. It is true that these regulators do not as a rule operate by threatening legal action in the event of non-compliance by the industry they regulate, and there has so far been no really significant litigation in the field of utility regulation. Nevertheless, the regulators work on a statutory basis and are able to cite the powers by virtue of which they make proposals, require information, or take decisions in relation to the industries regulated. The relationship is a very different one from that which evolved years ago between the old nationalised corporations and the controlling ministries in which the power to issue directives more or less atrophied and was replaced by private negotiations in which all too often ministerial political preferences prevailed over economic realism. It could also be held that regulation has analogies in other areas of standard setting and inspection, for example in education and health care. The idea of measuring performance and of setting targets has caught on, and thus become an integral part of the movement designed to make standards explicit and to test whether they are achieved or not. Whether these developments are good or bad is not a matter to be resolved here. It is enough to stress that they do represent a marked shift away from the informal and discretionary tradition in British administration towards a somewhat uneasy mix of legalism and market forces, even though the latter may often simply be simulated.

The other indicator of change worth special mention is the narrowing of ministerial discretions in sensitive areas affecting the freedom and rights of the individual. This has particularly affected the Home Secretary in whom many discretionary powers were and still are vested, but it extends to other spheres like education and mental health. Notwithstanding the desire of some recent Home Secretaries, notably Michael Howard, to assert their powers and responsibilities, the tendency has been towards putting some of the Home Secretary's discretions 'into commission' by introducing or reinforcing intermediate advisory or adjudicatory bodies. Appeals against refusal of entry to would-be immigrants and against deportation are no longer in the exclusive prerogative sphere of the Home Secretary and indeed have been subject to immigration adjudicators and an appeal tribunal since the early 1970s. His powers in relation to the release of prisoners have in effect passed over to parole boards (originating in legislation of 1967 and now also subject to judicial scrutiny), though in respect of remissions for groups of prisoners the Home Secretary retains the powers he has always had. A Police Complaints Authority handles the investigation of complaints against the police, though this does not stand in the way of the government instituting special inquiries, perhaps by a judge, into cases held to be of exceptional seriousness.[27] Very recently, too, the House of Lords has ruled that the Home Secretary has no power to vary the so-called prison tariff once this has been set by him or a preceding Home Secretary, while the European Court of Human Rights ruled in 1996 that a court and not the Home Secretary should determine the minimum period of imprisonment to be served by those under 18 years of age and detained at Her Majesty's pleasure. This narrowing of the Home Secretary's discretionary remit necessarily brings with it a narrowing of the range of matters for which he can be called to account in the House of Commons: more reliance on independent, judicial-type bodies necessarily reduces the scope of old-style political accountability.

JUDICIALISATION IN PERSPECTIVE

Judges and the courts, legal proceedings and judgments, formal regulation and independent adjudication undoubtedly figure more prominently in the manner in which Britain is governed and manages its affairs than they did 30 or 40 years ago. Moreover, the judicial resolution of disputes and resort to judges for objective and impartial opinions on matters of public concern and policy are looked on more favourably than they were in the past, and especially by those who see themselves as constitutional reformers. But is this trend to be seen chiefly as a reaction to the recent period of executive hyper-activity or does it express wider social developments which point towards a more judicialised mode of politics?

When trying to answer this question it is worth noting initially that there is nothing peculiar to Britain about the greater prominence of the judicial resolution of conflict and controversy in the public sphere. Continental Europe provides many examples of the same trend and in some countries, notably Germany, the judicial determination of issues which are fundamentally political and affect public policy has advanced so far that there are voices now raised questioning whether this makes for a healthy democratic public life. The presence there of a layer of specialised constitutional law interpretation does, of course, make for a degree of judicial intervention in the sphere of public decisions, including legislation, for which there is no real analogy in Britain and the German constitution itself legitimates this.[28] The EU also has many of the characteristics of a juridical construction and in its day-to-day operations often has to proceed on the basis of and by means of formal legal instruments, notwithstanding the fact that most decisions are the outcome of continuous political bargaining. The effects of this are manifest in Britain as in other member states in all spheres where Community law applies with direct effect and in varying degrees too where directives are to be implemented. Beyond the EU there are many states in eastern Europe which for obvious reasons are seeking to strengthen the judicial power as a safeguard against any renewal of the abuses of the past, whilst outside Europe there are many other notable examples of an increasing reliance on the judicial review of executive action and of a readiness to bring into force bills or charters of human rights enforceable through the courts. Canada, Australia, New Zealand and South Africa all exemplify this. And always exerting a powerful influence on the Common law world is the United States, with its uniquely long tradition of constitutional law interpretation. It is hard to underestimate the influence of American legal philosophy as well as judicial interpretation in furthering, both in the United States and in Britain, what we may call a rights-based political culture. Finally, there is more widely still the impact of the human rights movement throughout the international community and within the framework of the United Nations. This has not only influenced conceptions of the judicial role in Britain: there has too been a major contribution to the propagation of universalistic notions of human rights by British jurists, political and social theorists, politicians and moral crusaders which has helped to shape opinion both at home and abroad.

Rights are generally viewed as claims that the individual is entitled to make both against other persons and against public institutions. The possession of such rights is regarded as an essential part of the modern understanding of freedom. But democracy is also held to entail equal freedom for all and this in turn must mean equal rights and claims for all. In an articulate rights-based culture it is inevitable that the role of law will be enhanced and that people will come to look more often to lawyers, judges and formal institutions of adjudication for the settlement of many of the claims they raise. To a large extent the political and administrative authorities are as a result displaced: they cannot fulfil the demand for the authoritative arbitration of claims which arises in such conditions.

The policy changes of the 1980s and after as well as the style and approach of the governments of that period stimulated considerable political controversy, one by-product of which was an increasing resort to judicial challenge. But the response of the courts was variable – sometimes they administered checks to the claims of the executive, sometimes they rejected outright the pleas put to them by litigants attacking government actions. The overall rise in judicial arbitrations in many fields appears to owe far more to broader shifts in social attitudes and expectations which have been referred to above than to particular decisions and policies of the government of the day. It signals a declining willingness to trust those in power at all levels and a reluctance to recognise that there may be benefits in trusting officeholders to exercise prudently a reasonable degree of discretion. Perhaps it even points to a shift away from the traditional British preference for treating every case on its merits. What is, however, beyond doubt is that if this movement towards a more rule-bound mode of government persists in Britain, it will represent a sharper break with the past and a more profound change in political attitudes than it has done in many other countries. For Britain is virtually unique in having sustained to this day an informal and uncodified constitution which has kept at bay the excesses of legalism. This has been possible because of the relationship of accountability between ministers and Parliament and the existence of public services committed to procedural fairness and political neutrality. The advance of judicialisation looks like being but one of several indicators of the erosion of these traditional political values and practices.

There are many signs now pointing towards further reinforcement of formalised approaches to the resolution of conflict and the arbitration of claims. The new government elected to office on 1 May 1997 promised to incorporate the European Convention of Human Rights into British law and has now put forward legislative proposals to that end. But such a measure seems likely to strengthen the role of judges, though it also carries with it the risk of conflict with Parliament at some stage down the road. Freedom of information legislation is also promised, and that too can be expected to prompt legal disputes about what information can be accessed and what not. Devolution also holds out the prospect of arguments about the interpretation of devolved powers which at any rate in the case of Scotland might call for judicial resolution. It is also possible that were a settlement to be reached in Northern Ireland, that too might call for more frequent recourse to judicial arbitration than ever occurred in the past relationship of Northern Ireland with Great Britain.[29]

These prospects suggest questions for the future. If the role of judges as final decision makers, officers of the state who provide conclusive answers, is to be further enhanced, what happens to traditional British notions of democracy and responsible government? The wider the judicial role becomes, the more likely it is that the judges will be drawn into determining political questions, no matter what intellectual contortions may be performed in trying to deny this. Yet, it is the accountability of elected politicians that has been at the heart of modern British theories of government, and it is that theory, along with the authority of Parliament, which will be in competition with the judicialisation of politics. It is still not clear that the British people will in the majority prefer to put more faith in irremovable judges than in removable politicians. Moreover, there is also the paradox of freedom to be considered. More laws and rules, more higher norms and binding judicial interpretation may be called for by reformers in the name of equal rights and freedoms for all. But in the end an evolution in this direction may benefit a disparate range of minorities at the expense of diminished freedom and responsibility for the majority. Those like judges and members of tribunals who are not accountable to an electorate for their actions can also make mistakes, and that is all the more likely to happen when it is abstract principles that they apply.

NOTES

1. *Entick v Carrington* (1765) 19 State Trials 1030 laid down the principle that general warrants could not be used to justify searches or seizure of property.
2. It should be remembered, however, that the common law was never blindly prejudiced in favour of private property rights. In the field of freedom of contract, for example, the courts steadily imposed restraints on absolute freedom of contract in the course of the eighteenth and nineteenth centuries. See P.F. Atiyah, *The Rise and Fall of Freedom of Contract* (Oxford: OUP 1979). Furthermore, even Dicey complained of the way in which the common law judges went along with what he saw as a harmful collectivism driven forward by Liberal social-reform measures: see A.V. Dicey, *Lectures on the Relation between Law and Public Opinion in England during the 19th Century* (London: Macmillan 1905).
3. In the 1930s socialists like Laski and Cripps openly talked of the need for an enabling act to

override judicial objections to socialisation measures and of replacing reactionary judges. A far more benevolent critique of the social bias of the judiciary is offered years later by J.A.G.Griffith, *The Politics of the Judiciary* (London: Fontana 1977).

4. Only 20 years ago a Labour minister (Shirley Williams) presented amending legislation to Parliament in order to override the effects of the decision in *Secretary of State for Education and Science v. Tameside Metropolitan Borough* [1975], 2 *WLR* 641 which was at odds with her preference for comprehensive schooling. It seems doubtful whether today any Government would act so quickly and decisively to assert the supremacy of the Crown's ministers and Parliament on a major policy issue, though there are some recent examples of Government legislative action to reverse court decisions such as one on the payment of social security benefits to asylum seekers.

5. For valuable comments on this phase of judicial review and on Lord Reid's contribution in particular see J.A.G.Griffith, 'Judges and the Constitution, in Law, Society and Economy', in R.Rawlings (ed.) *Centenary Essays for the L.S.E. 1895–1995* (Oxford: OUP 1997).

6. There can be no doubt that the changes made in employment law from 1980 onwards have had a profound influence on attitudes towards legally enforceable frameworks of social regulation. Such methods have gained much wider acceptance than they used to enjoy, and in employment disputes industrial tribunals now play a large part in resolving disputes, with some contribution from the courts on points of law.

7. The figures quoted are taken from the Annual Reports of Judicial Statistics issued by the Lord Chancellor's Department. They must, however, be treated with great caution. The number of cases in which judicial review is actually granted remains very small. In 1993 it appears to have been 133 in civil matters and 127 in criminal cases. The statistics understandably throw no light on the qualititative aspects of the successful applications for review and they are complicated by the fact that many applications simply fall by the wayside and are abandoned for one reason or another. For detailed analysis of the procedures involved see [1992] *Public Law* 102.

8. In *M v. Home Office* [1993] 3 *All ER* 537 (HL) the judges concluded that a minister could be in contempt of court. This followed an executive decision to go ahead with a deportation in breach of assurances given to the court.

9. *R v. Secretary of State for Foreign and Commonwealth Affairs ex parte World Development Movement Ltd* [1995] 1 *WLR* 386.

10. [1993] *AC* 543.

11. *Bromley LBC v. Greater London Council* [1983] 1 *AC* 768. Lord Denning's judgement was subsequently upheld by the Law Lords.

12. The process of subjecting the police to much closer checks on how they proceed owes, however, more to legislation than to judicial decisions, e.g. the Police and Criminal Evidence Act 1984.

13. One of the best known and more influential of these constitutional reform groups has been Charter 88. This was founded in 1988 and has campaigned extensively for a 'modernised' constitution in tune with contemporary ideas of democracy. The protagonists of such an approach rarely take account of the tensions between the advocacy of more democracy on the one hand and an emphasis on formal rights and their enforcement through the courts on the other.

14. Among relevant articles or lectures of the judges mentioned are Sir H.Woolf, 'Judicial Review: a possible programme for reform' [1992] *Public Law* 221; Sir John Laws, 'Judicial Remedies and the Constitution' [1994] 57 *MLR* 213; Sir Stephen Sedley, 'Human Rights: a Twenty-First Century Agenda' [1995] *Public Law* 386. For a robust rebuttal of such judicial arguments tending to place the courts above Parliament see Griffith, in Rawlings (note 5).

15. Sir Leslie (now Lord) Scarman, *English Law: the New Dimension* (London: Stevens 1974).

16. The report on the conduct of Mr Profumo was published as Lord Denning's Report, Cmnd 2152, 1963.

17. Lord Justice Scarman's report was published as *The Brixton Disorders 10–12 April 1981*, Cmnd 8427.

18. *Report of the Inquiry into the Export of Defence Equipment and Dual-Use Goods to Iraq and Related Prosecutions*, presented to the House of Commons and published by HMSO, Feb. 1996 (5 Vols).

19. *R v. Secretary of State for Transport, ex p. Factortame Ltd. (No.2)* [1991] 1 *AC* 603.

20. The Home Office made a major contribution to drafting the Convention. Certainly in 1950 nobody in official circles in Britain ever dreamt of the prospect that the Convention would be used in litigation to plead the rights of transvestites or to challenge the validity of courts-martial. Remarkably there is no way of amending the Convention and no means of questioning the court's judgements other than by refusing to pay attention to them.

21. Such statistics as are available from the ECHR indicate that several other countries, e.g. Sweden, the Netherlands, Italy and France also contribute significantly to the number of successful appeals to it. Source: European Court of Human Rights: Survey of Activities 1992 and 1993.

22. Whether such a sop to parliamentary sovereignty will be worth much in practice appears to be doubtful. Parliament could certainly reserve its ultimate authority in relation to the effects of a purely domestic bill of rights, but it is more doubtful whether this is practicable so long as Britain remains a signatory of the European Convention. It is more likely that even after incorporation of the Convention into British law appeals to Strasbourg will have to remain possible and that some litigants will take advantage of that facility.

23. The banning of trade unions at GCHQ (General Communications Headquarters) in Cheltenham in 1984 (revoked by the new government in 1997) was one example of a political decision which did lead directly to legal challenge before the British courts and ultimately in Strasbourg too. On this occasion the Government's action was upheld as being within its powers to act on grounds of national security: *R v. Secretary of State for Foreign and Commonwealth Affairs ex p. Council of Civil Service Unions* [1985] *IRLR* 28.

24. Comparisons with the past must, however, always be qualified. The Wilson government 1966–70 had many ambitious legislative projects, and so had the Heath government 1970–74. But both suffered from notable setbacks, whereas Mrs Thatcher on the whole, did not.

25. The establishment of over 100 agencies stems from the report entitled *Improving Management in Government: The Next Steps* (HMSO 1988) which was adopted by the Thatcher government as the blueprint for a far-reaching programme of administrative deconcentration.

26. The Crown Prosecuting Service was established in 1986. It has been beset by many problems, some of them stemming from British laws of evidence, some from uneasy relations with the police authorities. As a result its operation continues to prompt critical comment.

27. The Police Complaints Authority was established under the Police and Criminal Evidence Act 1984 and replaced the Police Complaints Board set up in 1976.

28. For comments on the German predilection for judicial rulings and enforceable conclusions see Nevil Johnson, 'Law as Articulation of the State in Western Germany', *West European Politics* 1/2 (May 1978) pp.177–92.

29. During the 50 years of a devolved government and legislature in Northern Ireland there was hardly any recourse to such judicial remedies as might have been available against the abuse of power by the authorities there.

The Periphery and its Paradoxes

WILLIAM L. MILLER

PARADOX: seemingly self-contradictory or absurd but perhaps
well founded
Oxford English Dictionary

Paradox on the periphery? Manifest absurdities with a grain, or more than a grain, of truth in them? There is something about the territorial that brings out an abundance of the absurd. Focusing especially, but not exclusively, on Scotland, I want to consider the following propositions:

- The 'West Lothian question' is *not* a question.
- Tax-raising powers do not affect the ability to raise taxes.
- Devolution to Northern Ireland is no model for Scotland or Wales.
- Scottish Conservatives, Liberals and Labour all oppose the electoral system that favours them.
- Electoral systems for devolved assemblies have no implications for Westminster.
- It is possible to define moral limits to separatism.
- Sovereignty can be asserted by giving it away.
- Subsidiarity can be resisted by demanding it.
- Unionists are nationalists.
- Tories and Labour can swap sides without a pause in their fight.
- Devolution could only be achieved in conditions that made it unnecessary.
- Scottish political culture is *not* distinctive.
- Devolution is *not* an important issue for the Scottish public.
- People in Scotland do *not* identify themselves as Scottish.

It would be easy to extend the list, but this set of propositions ranges widely across matters that concern the constitution, all the major parties and the public. They are all – I hope – manifestly absurd. My purpose will be to search for grains of truth amidst this heap of absurdities.

THE 'WEST LOTHIAN QUESTION' IS NOT A QUESTION

The 'West Lothian Question' could be dismissed as a mere technicality were it not for its important consequences. It will be for ever associated with Tam Dalyell, formerly the Labour MP for the West Lothian constituency, now abolished. Dalyell had held the seat for Labour against a strong nationalist challenge at the 1962 West Lothian by-election. When Labour switched tack in 1974 towards supporting proposals for a devolved assembly, Dalyell found it hard, indeed impossible, to concede anything at all to nationalist sentiment. He was the only Scottish Labour MP to vote against the principle of the 1978 Scotland Bill (at the second reading, in November 1977) and he was a leading campaigner for a 'no' vote in the 1979 referendum on his party's devolution plan. Dalyell made a habit of reiterating for years on end a 'question' to which, he implied, there could be no acceptable answer. Later, he used his 'Belgrano Question' to devastating effect against Mrs Thatcher.[1] But in the late 1970s he went on to coin the 'West Lothian Question': why, he asked, should English MPs in the British Parliament be unable to ask questions or vote on matters such as Scottish education which would be devolved to a Scottish Assembly, while Scottish MPs would still be able to vote on purely English matters, such as school education in England? It might be called the 'Catalan Question' since it describes the position of Catalan MPs in the Spanish national assembly, except that it does not seem to trouble the Spaniards.

Two decades later, as the 1997 election approached, the *Scotsman*, formerly an extreme pro-devolution newspaper, attempted to resurrect this question as an important election issue. That was a paradox in itself, but one that is easily explained: under new owners the paper had a team of right-wing editors headed by Andrew Neil, recently released from other more pressing duties by Rupert Murdoch. In the larger excitements of Tory sleaze and prospects of a Blair government, the *Scotsman* failed to draw much attention to the West Lothian Question before the 1997 election. But afterwards, as Blair moved to implement devolution, the question was raised again and supplemented by a 'Garry McAllister question'. As Lawrence Donegan wrote in the *Guardian:* 'Donald Dewar [the new Labour Secretary of State for Scotland] is a formidable debater, but not even his nimble mind has found a plausible answer to the West Lothian Question in all these years. God knows how he will explain away the fact that Garry McAllister, Scotland's football captain, will not have a vote in the referendum – while Paul Gascoigne will.'[2]

No doubt Dalyell and Donegan would claim that the West Lothian and Garry McAllister questions were examples of paradox on the periphery, but they would be wrong. The real paradox here is not to be found in the content of the questions, but in the way that they were repeatedly asked, with eternal innocence, long after they had been effectively answered – like the bored toddler's repeated whine to its mother: 'but why?', no matter how often it is told the answer.

The West Lothian question raises a problem that does not exist either in principle or in practice, yet it has drawn intermittent attention for two decades. That is a real paradox. Let us consider the principle first. Tony Blair gave the short answer to the question in two words, during the 1997 election campaign: 'parliamentary sovereignty'. The central principle of British government is that proclaimed in 1885 by A.V. Dicey, Professor of Law at Oxford: 'the one fundamental dogma of English [sic] constitutional law [is] the absolute legislative sovereignty or despotism of Crown in Parliament'.[3] It does not matter that Dicey reneged on that principle when he found Parliament passing legislation (over Irish Home Rule) that he personally disliked. Nor does it matter that a Scottish judge in the 1950s gave a somewhat equivocal judgment that has been interpreted, by some but not by all, as a rejection of the principle of parliamentary sovereignty within Scotland. In so far as it makes any sense at all to talk about the principles of British government, the absolute sovereignty of Parliament is central to those principles.

Devolution has to be seen in that context. Before or after devolution the British parliament would be free to investigate, and legislate on, all matters including those devolved to a Scottish assembly. That is why Enoch Powell so memorably, and correctly, noted that 'power devolved is power retained': devolution is always conditional, always subject to override or reversal at any time. The zealots of the Scottish Constitutional Convention – such as Canon Kenyon Wright, might not agree, but they do seek to overthrow the principles of British constitutional practice, not to define them. Nothing, therefore, could prevent English MPs from voting on Scottish affairs except the House of Commons itself – dominated, as it always is, by English MPs. Only the British Parliament could prevent itself from dealing with Scottish affairs. Moreover, no self-imposed restriction by today's Parliament could bind its successor.

And suppose Parliament chose, collectively, not to interfere in devolved matters. Would devolution give a Scottish MP more power than at present? Certainly not. All that would happen is that an Essex MP would not be called upon to legislate on internal Scottish affairs – and there is no evidence that such MPs – other than Scottish *émigrés* such as Sir Teddy Taylor[4] – have ever found internal Scottish affairs anything other than a bore. Since they have never enjoyed having to think through the details of

Scottish affairs in the past, all they would lose would be an unwelcome chore. What is more, surveys show that the English electorate supports devolution for Scotland,[5] which means that the English electorage does not wish to retain day-to-day control of Scottish affairs.

So much for the principle. As for practice: British politics is in practice, party politics. In party terms, the West Lothian Question should be reformulated as: why should an English majority for one party be overturned – on purely English affairs – by an opposing majority in Scotland (or in Scotland and Wales)? This is indeed a practical question, but it turns out to be of no practical importance. It never happens. Since there are eight times as many English MPs as Scottish, even a strong imbalance towards one party in Scotland can do no more than tip the scales against the balance of party support in England. When that happens the overall party majority in parliament is so low that few radical reforms are enacted, either in England or elsewhere. Instead, such parliaments are short and feeble and the only important question at issue is the date of the next election – which is not a purely English question. The only three elections at which a Conservative lead amongst English MPs was outweighed by a Labour lead among (the much smaller number of) Scottish MPs were in 1950, 1964 and February 1974 – and all were followed by very short parliaments.

Moreover, the focus upon English MPs alone is misguided: apart from British and Scottish legislation, most of the remainder applies, not to England alone, but to England and Wales taken together. No one could dispute the right of Welsh MPs to participate in votes about that. The proposed Welsh assembly will not even be a legislative body! In 1950 and 1964, Labour was ahead in 'England and Wales' taken together; indeed Labour was not merely ahead, but had an overall majority in 'England and Wales' on both occasions. The only occasion in history when a Conservative lead in 'England and Wales' was overturned by the presence of Scottish MPs was February 1974. And it is worth noting that in February 1974, the overall result was a hung parliament in which no party – Labour or Conservative – gained an overall majority.

It needs some emphasis: there has never ever been any occasion on which the Conservatives have had a majority in 'England and Wales', but Labour has had a majority in parliament as a whole. In short the situation postulated in the West Lothian question, always unlikely because there are so few Scottish MPs, has never ever occurred in practice. And, needless to say, a Labour majority in England has never been overturned by Scottish MPs.

The more recent Garry McAllister Question has an even simpler answer. The politics of devolution and even of nationalism in Scotland are firmly 'civic' rather than 'ethnic'. This terminology may be more familiar in discussions of, for example, Ukrainian politics where careful distinctions are made between 'Ukrainians' (ethnic Ukrainians) and 'the people of Ukraine' (all those, including ethnic Russians, who live in Ukraine and are, therefore, entitled to be citizens). This 'civic nationalism' contrasts with the 'ethnic nationalism' of Estonia where many long-time residents are excluded from full rights of citizenship. The concepts are very familiar in Scottish politics and the terminology is unknown simply because no significant nationalist group in current Scottish politics advocates 'ethnic nationalism'. Even the Scottish National Party (SNP) – and perhaps we should say, *especially* the SNP – emphasises the concept of 'those who live in Scotland' rather than an ethnic/racial definition based on birth or blood. In Scottish politics, only the Unionists, between the wars, came close to the ethnic nationalist perspective, in their antagonism towards Irish Catholic immigrants.[6] The paradox that Unionists can be extreme nationalists is one to which we must return later. For now it is enough to note that Scottish nationalism could be ethnic or it could be civic but, in fact, it is very firmly civic in contemporary Scottish politics. The idea that Scots *by blood* should vote in a referendum on civic devolution *by territory* is so self-contradictory that it has never occurred to anyone other than the opponents of devolution. A referendum by blood would only make logical sense in terms of devolution by blood – along the lines advocated in the Habsburg Empire and practised in the Ottoman.

Both these superficially weighty and solemn constitutional 'questions' are not questions at all, but frivolous debating slogans. The answers are obvious, simple and straightforward – but are unwanted and ignored by those who reiterate the questions.

TAX-RAISING POWERS DO NOT AFFECT THE ABILITY TO RAISE TAXES

There was great indignation in the Scottish press and among the enthusiasts of the Scottish Constitutional Convention when Tony Blair decided that Labour should propose a two-question referendum on devolution in Scotland.[7] The idea of a referendum to underline the Scottish people's support for the creation of a devolved Scottish assembly was already accepted by the devo-enthusiasts. But the idea of a referendum to confirm the people's support for giving that assembly tax-raising powers was not. Allegations of 'betrayal', however, raised the question of who had been betrayed by Blair's change of tactics – hardly the Scottish people who could not be betrayed by being consulted, but perhaps the Convention enthusiasts who felt that they had struck a deal with Labour which did not have public support and which might therefore be overturned in the referendum. It was certainly remarkable that some of those who so stridently asserted the Scottish people's 'claim of right' wanted to prevent the people from exercising it.

That was not the only irony about tax powers for a Scottish assembly however. Labour – desperate in Scotland no less than in England to avoid the label of a 'high tax' party – had already promised that it would not use tax-raising (strictly 'tax-varying') powers in the first term of a Scottish assembly, even if they were granted; and it would not use them in any subsequent term of that parliament without a prior manifesto commitment to raise an additional Scottish tax. Moreover, British Social Attitudes Surveys throughout the 1980s had shown that, in the long term, Scots were no more favourable to high taxes than were the English.[8] That would make it quite dangerous politically for Labour to give a manifesto commitment to an additional tax in Scotland. Scottish Conservatives would revel in such an election campaign for a Scottish assembly. So formal powers to increase income tax in Scotland looked little more than symbolic: they would certainly not be used in the near future, and they were none too likely to be used at all. In short, the grant of formal tax-powers would not give a Scottish assembly a real ability to raise taxes in practice.

Conversely, however, the failure to grant formal tax-powers would not prevent a Scottish assembly from raising taxes in practice. The previous devolution scheme, in 1978/79, had been criticised for lacking the tax powers necessary to inculcate a sense of responsibility. It would, said critics, spend its time complaining about the lack of resources provided by Westminster. Tax-raising powers would force it to confront its own electorate with the choice between better public services and lower taxes. Oddly enough, it was the same people – Conservatives mainly – who had criticised the 1979 scheme for encouraging irresponsibility by omitting tax powers as now criticised the 1997 scheme for including what they dubbed a 'tartan tax'. Paradoxically, however, the omission of formal tax-powers would not, in fact, have prevented the proposed 1979 Scottish assembly from varying taxes, nor would it do so in 1997. The formal tax-varying powers proposed in 1997 – up to three pence on the basic rate of income tax – could raise relatively little revenue at most, compared to the size of the block grant for public services (including subsidies to local government) that would be provided by Westminster. By varying the amount of that block grant that was, in fact, used to pay for local government and other public services in Scotland, the Scottish assembly– under the 1979 or 1997 proposals – could effectively determine the level of local council taxation ('rates', 'poll tax', or 'council tax') and/or the extent of consumer charges for public services in Scotland, both of which would give it an enormous ability to raise or lower taxes (and/or charges) on individuals in Scotland. Whether or not it would choose to use that ability is another matter, but one that is exactly equivalent to the question of whether or not it would choose to use formal tax-powers. Indeed, since use of these informal powers would be less visible to the electorate than its use of formal powers to vary income tax, we might suspect that the informal powers would be more usable than the formal.

There is a final irony about a Scottish assembly's tax-powers: under proportional representation, Labour would not dominate the assembly and therefore would be less able to guarantee that formal tax-powers would not be used. Both the Liberals and the SNP were in favour of higher taxes and together could expect over a third of the seats in a Scottish assembly. If proportional representation also eroded Labour discipline – by making it more possible for a left-wing breakaway group to retain their seats without official Labour backing – then the Labour leadership might *not be able* to guarantee that formal tax-raising powers would not be used – unless, of course, they could rely on wholehearted co-operation from the Conservatives. A Machiavellian Conservative Party might campaign strongly in the media against higher taxes while mysteriously ensuring sufficient absences from Parliament to allow a coalition of Liberals, SNP and disloyal Labour leftists to impose a 'tartan tax' as they had always warned might happen. Or is that too Machiavellian even for the Conservative Party? Those who remember the tactics used by Labour to postpone the parliamentary boundary revision in November 1969 would not think so.[9]

DEVOLUTION TO NORTHERN IRELAND IS NO MODEL FOR SCOTLAND OR WALES

From 1922 to 1972, Northern Ireland was governed by a devolved assembly in Stormont. During that half century, British parliaments and governments interpreted the degree of autonomy implied by devolution very generously, until their patience eventually snapped. Since 1972 there have been intermittent attempts to restore some form of devolved government in Northern Ireland. Occasionally Scottish devolutionists asked why a British government should seek to restore devolution to Northern Ireland where consensual support for it was lacking, yet oppose devolution for Scotland where it was available. Prime Minister John Major in particular kept warning that devolution for Scotland would mean the end of the United

Kingdom, yet he remained an advocate of devolution for Northern Ireland, provided some system acceptable to both religious communities in Northern Ireland could be found.

This certainly seems inconsistent. But is it? Both the infrequency and the muted tone of the Scottish devolutionists' references to Northern Ireland suggests not. Both sides of the devolution argument in Great Britain seem tacitly agreed that Northern Ireland is a place apart, not 'really' part of the United Kingdom, despite its formal constitutional position and both tacitly agree that it can provide no model for government anywhere else. No part of Great Britain wants to be like Northern Ireland. Psychologically it is a place for which most British politicians might have tremendous sympathy but a *place* with which they cannot identify, even if they can identify with a faction within it. Occasionally that perspective is revealed in laws such as the Prevention of Terrorism Act which allowed exclusion of Northern Irish residents from mainland Britain, and even deportation from mainland Britain to Northern Ireland – 'internal exile' as it would have been called in Tsarist Russia. At an administrative level, the experience of the Stormont system does provide a great deal of useful guidance about the details of how to run a devolved system. But at a political level it provides a great deal of useful guidance about the details of how *not* to run a devolved system, and even discourages support for the principle itself. The proposition: 'we must be treated like Northern Ireland' is too close for comfort to the proposition: 'we are like Northern Ireland' or 'we could be like Northern Ireland'. Psychologically, Northern Irish devolution is associated with division and violence. After such a long history of troubles, Northern Ireland can always serve as a warning, but never as a model.

FOR A SCOTTISH ASSEMBLY, THE CONSERVATIVES, LIBERALS AND LABOUR ALL OPPOSE THE ELECTORAL SYSTEM THAT FAVOURS THEM

It is almost axiomatic that parties prefer electoral systems that help themselves, or at least their close allies.[10] But for the proposed Scottish assembly, three of the main parties oppose the electoral system that works in their favour, and none of them do it primarily to help their allies.

The 1997 Scottish Conservative conference in Perth overwhelmingly backed the party leadership's opposition to devolution. Conservatives oppose the Scottish Constitutional Convention's plans for a Scottish assembly both in principle and in detail. That implies opposition to the proposal for a PR (proportional representation) electoral system and continued support for Britain's traditional FPTP (first-past-the-post)

TABLE 1
SEATS TO VOTES RATIOS IN SCOTLAND
(PERCENTAGE OF SEATS DIVIDED BY PERCENTAGE OF VOTES)

	Conservative	Labour	Liberal	SNP
1955	1.0	1.0	0.7	0
1959	0.9	1.2	0.3	0
1964	0.8	1.2	0.7	0
1966	0.8	1.3	1.0	0
1970	0.9	1.4	0.8	0.1
Feb. 1974	0.9	1.5	0.5	0.5
Oct. 1974	0.9	1.6	0.5	0.5
1979	1.0	1.5	0.5	0.2
1983	1.0	1.6	0.5	0.2
1987	0.6	1.6	0.7	0.3
1992	0.6	1.8	1.0	0.2
1997	0	1.7	1.1	0.4

system. Conversely Labour and the Liberals back PR for the devolved Scottish and Welsh assemblies. Yet in the 1997 general election, FPTP damaged the Conservatives and helped, not only Labour, but also the Scottish Liberals.

Under the classic 'two-and-a-half party system' that applied in Britain during the 1950s and 1960s, the two large parties might each expect to get almost half the seats in Britain with a vote of just under 45 per cent, while the Liberals could expect only a handful of seats for the remaining 10 per cent of votes. So, in that slightly idealised situation, the 'seats to votes ratio' (strictly the *per centage* of seats divided by the *per centage* of votes) would be just over 1.1 for the two main parties and close to nought for the Liberals. At the closely fought general election of 1964, for example, the seats to votes ratios in Britain as a whole were 1.1 for both the Conservatives and for Labour, but only 0.1 for the Liberals. The adverse ratio for the Liberals reflected both the size of their vote – only 11 per cent – and the fact that it was so evenly spread.

But that was not how the FPTP system operated in Scotland. In Scotland, the Liberal vote was even smaller than in Britain as a whole, but it was far more concentrated in a few rural constituencies. So the seats to votes ratio for Scottish Liberals was much better than for English Liberals (Table 1). In 1964 and 1970 it was almost as favourable as the seats to votes ratio for the Scottish Conservatives; and in 1966, 1987, 1992 and 1997 it was better than for the Scottish Conservatives. Indeed, in 1992 it was as good as the Liberals could expect under a very proportionate PR system, and in 1966 and 1997 it was better than under a PR system. Conversely, the

Conservatives did not do well under FPTP in Scotland. They nearly always did worse than under PR. In terms of seats, PR would have given the Conservatives, five more seats in 1964, seven more seats in 1966, seven more in 1987, eight more in 1992 and 13 more in 1997 – massive under-representation in a country with only 72 seats in total. As the largest party, Labour got the typical FPTP bonus of seats in the late 1950s and early 1960s, with a seats to votes ratio of 1.2. But that rose steadily thereafter as the SNP vote rose and all other votes, including Labour's, declined: in broad terms, Labour lost votes without losing seats and, from 1974 onwards, Labour's seats to votes ratio was always over 1.5, reaching a peak of 1.8 in 1992.

Scottish Liberal policy was the least paradoxical. Despite their very small vote, they did about the same under FPTP as under PR and sometimes better, but never much better. So they can hold to their long established policy of PR – reflecting the advantages they would gain from PR in England – without hope of gain in Scotland, but without fear of much damage either.

The Conservatives were in a more difficult position. Under FPTP they were consistently under-represented in Scotland, and in 1997 they won no seats at all in Scotland – or in Wales – despite a substantial vote, far larger than that of the Liberals. They complained that they were given no seats on the Scottish or Welsh Grand Committees in parliament despite these substantial votes – the classic argument of the PR enthusiasts; and yet, they opposed PR in principle. We must assume that they feared the introduction of PR for Scottish and Welsh assemblies would be the first step on the slippery slope towards PR for the British Parliament, and they were willing to sacrifice their Scottish and Welsh representation in order to preserve their usual over-representation in England – though under the force of the 1997 Labour landslide, they were temporarily under-represented in England also. PR had been used in Northern Ireland in the 1920s and again after the abolition of Stormont. But as we have noted, Northern Ireland is never accepted as a model; so use of PR in Northern Ireland does not have the same implications for Britain as its use in Scotland or Wales.

But Labour's position was the most paradoxical. Although 56 of the 72 Scottish MPs elected in 1997 were Labour, the party has never achieved a majority of the vote at any general election in Scotland – unlike the Conservatives. It must assume that under PR it would have no majority in a Scottish assembly, or at most a slim majority at the mercy of maverick backbenchers. Given the experience of James Callaghan's government in the 1970s and John Major's in the 1990s that is not an obviously attractive prospect – neither for Labour nor for Scotland. So why should a dominant party seek to end that domination for ever? In political terms it seems almost suicidal. But the truth is that Labour was simply embarrassed by its success. The end of Labour domination in Scotland was the price that had to be paid for devolution. Without the assurance that Labour domination would be abolished there would be no consensus – no party consensus and no geographic consensus – in favour of devolution. The Liberals enthusiastically, the SNP grudgingly, and even the Conservatives under duress, could all accept a Scottish assembly provided they could be sure that it would not produce what eastern Europeans used to call a 'party-state' – a Scottish assembly that was just the Labour Party operating under a different name. Furthermore, Labour was identified with the central belt, the other parties with the north and south of Scotland and the Islands. Thus, these peripheral areas could accept a Scottish assembly more easily if they could be sure that it would not be totally dominated by the perspectives and interests of the central belt – as they feared it would be if it were dominated by Labour.

So, given its strong commitment to devolution, it is not absurd for Labour to adopt PR in order to destroy its own, embarrassing majority. The loss of its majority was not the cost of adopting the principle of PR. The principle, the objective, was to lose the majority by any means; PR was only the incidental means to that end. Whether the loss of its majority was a price worth paying for devolution is, of course, another matter entirely. Perhaps that was absurd.

ELECTORAL SYSTEMS FOR DEVOLVED ASSEMBLIES HAVE NO IMPLICATIONS FOR WESTMINSTER

The main reason why the Conservatives opposed a PR system for elections to the Scottish assembly, to the Welsh assembly – and, indeed, to the European Parliament – was that the use of PR for those elections could increase pressure to use PR for Westminster elections. No doubt Liberals supported PR in Scottish, Welsh and European elections for exactly the same reason. Politically they were both right; logically they were both wrong.

We all have a tendency to over-value consistency. It is a symptom of intellectual laziness; it avoids the need for fresh thinking. Once we have decided that PR is a good thing or a bad thing, we are inclined to apply that decision in all places, in all circumstances and for all purposes. Since the general public has little interest in politics, especially constitutional politics, they are unlikely to spend much time thinking through the subtleties of electoral systems. Consequently, to the uninterested public, acceptance of

PR for some elections would indeed be an argument in favour of using PR in other elections. The danger from the Conservative perspective, the opportunity from the Liberal perspective, is real.

But in logic, there is no reason why one system should be best for all elections. As its name suggests, PR is appropriate when the purpose is representation. Party proportionality does nor guarantee any other kind of proportionality of course, but it is at least a step in the right direction. As noted above, party-proportionality would automatically make a Scottish assembly more representative of geographic interests outside the central belt as well as within it. After a dispute between Labour which wanted gender proportionality written into the electoral system for a Scottish assembly, and the Liberals who were sympathetic to gender proportionality but did not want it written into the law, the two parties signed a solemn inter-party agreement in the presence of Scottish Convention officials, which committed them to use their internal party rules to ensure gender proportionality among their own MPs in the Scottish assembly. Although both conveniently ignored the Labour Party's founding principle of class proportionality, their aim was clearly to produce a Scottish assembly that was as representative as possible of the people of Scotland.

Representation, at least proportionate and perhaps disproportionately weighted in favour of minorities, is a good way of securing a broad consensus and reassuring minorities that their voices will be heard. But proportional representation has the almost inevitable corollary of government without a single-party majority in parliament – which, in the European tradition, means government dependent upon a shifting coalition and, in the British tradition, a minority government dependent upon the temporary disunity or tolerance of its enemies. That is not likely to be the best way of ensuring strong government capable of decisive action in an acute crisis – and indeed PR may be advocated precisely to avoid producing governments capable of decisive, precipitate and possibly tyrannical action. But for strong, decisive government we need a clear focus of responsibility rather than representation, a single voice rather than a cacophony of divergent opinions, a system that produces majority rather than minority or coalition governments.[11] The (democratic electoral) systems that produce such governments are FPTP, high thresholds or other mechanisms for 'gearing' parliamentary elections or, if all else fails, a presidential system.[12] (Non-democratic alternatives are always available, of course.)

So it is not really inconsistent to advocate a system such as FPTP for the level of government that has to deal with the most dangerous and most acute crises, including military threats, but to advocate a PR system for other levels of government. Nor to advocate a system such as FPTP to produce strong government at the centre of power, and PR to deliberately produce weak, naturally subordinate government at other levels. Nor, more concretely, is it inconsistent to advocate FPTP for state government and PR both for subordinate, devolved assemblies and for European level institutions. At least a combination of strong and weak assemblies or parliaments at different levels may help to avoid deadlock between two equally powerful, decisive and ambitious levels of government – the kind of 'war of the laws' between the USSR and its constituent republics that led to its break-up. Of course, different people have different objectives. But for those who want to concentrate power at the state level, and keep both Europe and the regions in a subordinate position, it is at least not inconsistent to advocate PR for Europe and the regions, along with FPTP or a President for state-level government.

It is, however, either a paradox or a real absurdity to advocate both PR and presidentialism – which are polar opposites – for the same level of regional government in different parts of the state. The 1997 Labour Party manifesto advocates PR for a Scottish assembly and a directly elected 'mayor' (a 'president' on a regional scale) for London.[13] The case for directly elected mayors in local or regional government is to give a focus not just for decision making, but also for popular identity and enthusiasm. No doubt they would do just that. Whether desirable or not, it would be fairly safe to have directly elected mayors in local government because they would represent populations too small and institutions too weak to challenge central government. A directly elected mayor for a region such as London would have much more 'clout', however, and would also be seen as an argument for a directly elected prime minister (or president) for the regional assembly in Scotland. And just as PR for a Scottish assembly is well-designed to prevent a concentration of power in Scotland that might alarm minorities within Scotland and challenge central government outwith Scotland, the direct election of a Scottish chief executive and chief spokesperson would do the opposite: the combination of devolution and a directly elected head of Scottish government could be explosive – within Scotland, and within Britain.

IT IS POSSIBLE TO DEFINE MORAL LIMITS TO SEPARATISM

This proposition is manifestly absurd. Separatists tie themselves in knots in their search for moral criteria to limit the process of separation. They may impose their will by force of arms but not by force of argument or reason. Tomas Masaryk was always better at explaining why Czechoslovakia should separate from the Habsburg empire than he was at explaining why

Slovakia should not separate from Bohemia and Moravia, or the ethnic Sudeten Germans from the Czechs of Bohemia. And in practice also, one assertion of difference, one concession of separation, prompts others to think in similar terms. In our own times, Georgia's exit from the Soviet Union encouraged Abkhazia's exit from Georgia; Ukraine's separation from the Soviet Union encouraged Crimean demands for independence from Ukraine or reunion with Russia – just as the prospect of Ireland's separation from the United Kingdom at the turn of the century stimulated Ulstermen to demand separation from Ireland and (continued) union with Britain, and then further separation of three of the nine counties of Ulster from the rest of Ulster – to leave only what some still call 'the six counties' in union with Britain.

Scotland is not totally immune from this vortex of separatism. Except in the eyes of romantic nationalists it is not united. Many outside the central belt look upon Glasgow and Edinburgh with almost as much suspicion as they do London. In the 1979 referendum on a Scottish assembly, there was a majority, albeit a narrow majority, in favour of a Scottish assembly throughout the central belt; but only 40 per cent support in the south of Scotland – the Borders, Dumfries and Galloway regions; and less than 28 per cent support in Orkney and Shetland. At that time, there were serious threats that Orkney and Shetland would defect from Scotland and establish a new direct relationship with London, taking their claims to North Sea oil with them.

In the event, even devolution, let alone Scottish separation, did not happen in 1979 and the threats from the islands never materialised. But the problem of principle remains: how, except by an appeal to force, perhaps disguised as an appeal to numbers, could separation be claimed for Scotland but denied for Orkney and Shetland? Or to southern Scotland? A *Scotsman*/ICM poll in mid-1997 purported to show that Orkney and Shetland were now reconciled to the prospect of devolution for Scotland. In fact, what it showed was that the mood in 1997 was far more favourable towards Scottish devolution than it had been in 1979. Consequently, there was a majority even in Orkney and Shetland for a 'yes' vote in the 1997 referendum. But the ICM poll showed that support for devolution remained much less in Orkney and Shetland than on the mainland and uniquely vulnerable to a general swing against Scottish devolution: what was close to unanimous support in parts of Scotland was clearly very ambiguous and conditional in the islands. Ominously, delegations of Orkney and Shetland councillors were to be seen, in mid-1997, visiting the Isle of Man – which comes under the Crown, but is not a part of the UK, nor even a part of the EU – to discuss the advantages of alternative constitutional arrangements.

On purely moral grounds it would be difficult to claim a right of separation for Scotland but deny it to Orkney and Shetland. In principle, the proposition that it is possible to define moral limits to separatism is indeed not a paradox, but an absurdity. The paradox is that the unanswerable question is of no importance unless it is seriously raised in practice. The SNP was very sensitive to the threat of breakaways from Scotland. It claimed to sympathise with all small communities even within Scotland – though the word 'nation' could hardly be conceded to the islanders – and it tried to accommodate Orkney and Shetland as far as possible. Similarly, Labour and the Liberals in the Constitutional Convention tried to develop a geographic as well as a cross-party consensus on devolution by taking account of fears that a Scottish assembly might be dominated by the parties, interests and perspectives of the central belt. As a result, there is now a clear majority for a Scottish assembly even in the islands.

SOVEREIGNTY CAN BE ASSERTED BY GIVING IT AWAY

The SNP has always been troubled by the concept of independence. It claims independence as its core policy, its *raison d'être*. But on closer inspection, independence turns out to be a matter of degree rather than something qualitatively different from membership of a wider polity, not a single well-defined status but a range of possibilities. In the first few decades of the party's existence the national movement in Scotland – sometimes synonymous with the SNP, at others much wider than the SNP – was divided between those who sought independence from Britain and those who, in modern language, sought something closer to devolution. The mood varied from time to time but generally the dominant mood in the party was for complete independence from Britain. That remains the present policy.

The SNP was antagonistic towards Europe as well as towards Britain in the 1970s, quite properly earning adherents the title of 'separatists', but in the late 1980s, Jim Sillars persuaded the party to contest that allegation by adopting a new policy of 'independence within' rather than 'independence from'.[14] To be sure, the policy remained one of independence *from* Britain, but independence *within* Europe. But none the less there is something inherently paradoxical about a nationalist demand to be independent 'within' anything. There is an implied voluntarism in the notion of 'independence within' which does give some meaning to the independence element. Still, the essence of the policy is to obtain sovereignty in order to trade it away again. Of course, British politicians, other than the Eurosceptic right of the Tory Party, are quite explicit about their willingness to trade some part of British sovereignty in return for the advantages, to Britain, of membership in the European Union (EU). People exercise their freedom by joining clubs which restrict their freedom in some ways, but enlarge their opportunities in others. That is normal for individuals and normal for sovereign states. Like money, it is often better to spend sovereignty than to hoard it. Conceding sovereignty is not at all absurd for many politicians. National politicians are not necessarily, or even usually, nationalist politicians, but it does seem absurd for nationalist politicians to concede the one thing that motivates and inspires them.

Donald Dewar reacted to the nationalists' new policy of 'independence within Europe' by describing Labour's devolution policy as one of 'independence within the United Kingdom'. It is a lot less absurd for a pro-devolution politician to advocate 'independence within' than for a nationalist, since the pragmatic devolutionist is ideologically committed to the very unromantic notion of trading elements of sovereignty for practical benefits: sovereignty, as such, is not something to which they are greatly attached. The evidence from all the opinion polls is that the bulk of the Scottish public does not feel that it is being coerced into membership of the United Kingdom. Psychologically at least, there is, therefore, an implied voluntarism in the existing situation that makes it more equivalent to membership of a club rather than to subjection to an alien empire and this justifies the phrase 'independence within the United Kingdom' – certainly with devolution and perhaps without it – in the same way as voluntarism can be used to justify the concept of 'independence within Europe'.

SUBSIDIARITY CAN BE RESISTED BY DEMANDING IT

The concept of subsidiarity can be traced back at least to a 1931 encyclical of Pope Pius XI. He was concerned that the state was usurping the traditional role of the church: 'it is an injustice, a grave evil and a disturbance of the right order, for a larger and higher association to arrogate to itself functions which can be performed more efficiently by smaller and lower societies.'[15] His own most immediate experience, of course, was of Mussolini's Fascist state and his whole argument could be dismissed as special pleading on behalf of the Catholic Church.[16] But out of that immediate and, for him, very personal experience, has come the clear principle, however obscure in practice, that governmental functions should be exercised 'on as low a level as possible and as high a level as necessary'. There is some subtlety but no paradox, no obvious absurdity, in that principle. The paradox arises in practice, when people chose to support only very selective applications of the principle. Viewed from the perspective of

British central government, full application of the principle of subsidiarity implies retaining some powers at Westminster, but shifting others upwards where necessary and downwards where possible. And the inverse of subsidiarity is state centralism: concentrating powers at Westminster by taking them both from Europe and from local government or other subordinate institutions within Britain.

Reacting to 18 years of Conservative Party domination at Westminster, the Labour Party has adopted a relatively consistent policy towards subsidiarity by the 1990s – sympathetic both to greater powers for the European Union and to regional devolution and greater local government autonomy within Britain. We cannot be sure that they will remain so, once they become more accustomed to holding power at Westminster. But right-wing 'Eurosceptic' Conservatives, including Mrs Thatcher whether in or out of office, have consistently opposed the principle of subsidiarity. The essence of Thatcherite philosophy was freedom for the unorganised individual within a strong British state. With her colleagues and successors she attacked all the 'intermediate organisations' that stood between the individual and the state – the trade unions, the professional bodies, universities, churches and local governments. She abolished the Greater London Council and the English Metropolitan County Councils, took powers to dictate both the taxation and expenditure plans of each local authority, transferred functions from elected local councils to appointed quangos and private businesses, and opposed any moves towards devolution for Scotland, Wales or the English regions. In Europe, Thatcherites backed a Europe of so-called nation-states and opposed anything that could be interpreted as European federalism. There is nothing at all absurd in all of this: it is a consistent attempt to concentrate power at a single level by taking power from levels above and below the (British) state – the exact antithesis of subsidiarity. Pope Pius XI would have recognised this policy though it was not subsidiarity: it was the policy of Mussolini that he was attacking in his encyclical. The paradox is merely that some Tory Eurosceptics have used the term 'subsidiarity' to describe what is its exact antithesis: in the name of subsidiarity they have tried concentrate power in one institution at one level.[17]

The only sense in which Thatcher could claim to be on the same side as Pope Pius XI was that he, like her, supported the concept of devolution 'down to the level of the individual', as Mrs Thatcher called it when explaining why her devolution policy was better than Labour's, or down to the level of 'the family', as Pope Pius expressed it – though both claims should be treated with scepticism. Pope Pius used devolution to the family to defend the power of the Church; Thatcher's devolution to the individual was used to defend and extend the power of private business against, for example, municipal public enterprise.

UNIONISTS ARE NATIONALISTS

To suggest that Ulster Unionists are ethnic nationalists might not be very controversial. The claim by Lord Craigavon, Unionist Prime Minister of Northern Ireland in 1934, that 'we are a Protestant parliament and a Protestant state'[18] reflects the classic nationalist demand that the state should be coextensive with the ethnic group. (He was, it should be said, responding to the assertion that southern Ireland was a Catholic state.) That language, and even that sentiment, it not heard so often or so loudly now, but Ulster Unionists of various hues are still engaged in the politics of ethnic defence, if no longer the politics of ethnic supremacy. Their defence of Protestant interests and traditions is as nationalist as the Social Democrat and Labour Party's (SDLP's) defence of Catholic interests and traditions. Understandably, Ulster Unionist politicians would be uncomfortable with the term 'nationalist' – but then so is John Hume of the SDLP, always ready to point out that he got into Parliament by defeating the Nationalist candidate (with a capital N). To outside observers the problem of Northern Ireland is clearly that of 'two nationalisms in one country'.

It is more of a paradox to assert that British mainland Conservatives including Scottish Unionists are nationalists. Let us put aside the charge that many English Conservative MPs are English nationalists; there is a deeper paradox than that. Some of those who are committed to the concept of the United Kingdom – or even of Britain – are at heart true federalists who see the UK as a multinational state, a voluntary organisation linking together fundamentally different peoples and countries. Sir Malcolm Rifkind tended to use such language. Such people might once have been imperialists but never nationalists: they did not identify, nor seek to identify, state and nation.

However, many of those who are most strongly committed to the concept of Britain are truly nationalist in the classic sense that they *do* believe the state should be coextensive with the nation, but they view the British people as a single nation albeit a nation which encompasses some merely local traditions and historical heterogeneity – after all, what nation does not? Like the nationalists of the SNP, they are in favour of separating their nation from a larger state or, in the shape of the EU, a larger state in embryo, but they oppose the secession of any part of the national territory from their nation-state. Within their nation-state they are by instinct centralisers; they want a clear line of division between their nation-state and other states, but no significant divisions within their nation-state. They tend to demand freedom *for* their nation but oppose freedom *within* their nation. By these criteria there is little difference between the SNP and most Conservatives except that their 'imagined community', their idealised nation, is the 'Scottish nation' in one case and the 'British nation' in the other. It would be wrong to suggest that either the SNP or the Scottish Unionists are extreme nationalist parties, but they are equally obsessed with the national idea – to an extent that neither Labour with its class based traditions nor the Liberals with their individualistic ideology could ever be.

TORIES AND LABOUR CAN SWAP SIDES WITHOUT A PAUSE IN THEIR FIGHT

The workers, said Marx, had no country. The socialist and class-based Labour tradition emphasised the class enemy, not the national enemy, class solidarity not national solidarity. Labour might sympathise with the poorer regions of Britain, as it sympathised with the poorer classes, but nationalism or separatism – anything beyond a policy of regional equalisation – was at best a distraction from its main objectives and at worst in contradiction to its core beliefs. As the Scottish Conservatives rightly pointed out, the Attlee government's programme of nationalisation meant denationalisation for Scotland, as power was centralised in London. That was in keeping with Labour's ideology of equality and 'national standards' for citizens throughout Britain: indeed there is an inherent contradiction between the concepts of local autonomy and national standards. But as long as Scotland and Wales lagged behind England economically, Scottish and Welsh interests were consistent with a drive for equal 'national' standards throughout Britain. The English might reasonably complain, but the Scots and the Welsh could not.

Yet, complain they did, and it was the Conservative and Unionist Party that led the chorus of peripheral nationalist complaints, despite its name. Under the Attlee government, Scottish Unionists shared platforms with Scottish Nationalists: at the 1948 Paisley by-election they put forward a joint candidate – John MacCormick. A week before the 1950 general election, Winston Churchill told a Conservative and Unionist rally in Edinburgh that: [t]he principle of centralisation of government in Whitehall and Westminster is emphasised in a manner not hitherto experienced or contemplated in the [1707] Act of Union. ... I frankly admit that it raises new issues between our two nations ... I would never adopt the view that Scotland should be forced into the serfdom of socialism as a result of a vote in the House of Commons.' Churchill's respect for parliamentary sovereignty was as weak as Dicey's[19] or Hailsham's[20] when parliament displeased them! 'The socialist menace has advanced so far', Churchill continued 'as to entitle Scotland to further guarantees of national security and internal independence'.[21]

Under a later Labour government, Edward Heath made his 'Declaration of Perth' at the Scottish Conservative conference in May 1968. His rhetoric was not so inflammatory as Churchill's, but Heath committed his party to set up some form of Scottish assembly. To decide the details, he set up a Scottish Constitutional Committee chaired by former prime minister, Lord Home. It reported in 1970, advocating an elected Scottish Convention to take over work then done by various Scottish committees at Westminster. The 1970 Conservative manifesto committed the party to implement the plan if elected. Meanwhile, Willie Ross, the Labour Secretary of State for Scotland, led the attack on 'separatists': 'How can a separate government sitting in Edinburgh', he asked the Scottish Trades Union Congress Conference in 1968, 'put the squeeze on firms in the South East and Midlands and drive them up to Scotland?'[22] In 1970, the Labour manifesto declared it would 'reject separatism and also any separate legislative assembly'. So as late as 1970, the Conservatives were still on the devolutionist side, and Labour still against it. Indeed that remained true at the February election of 1974.

But Labour switched tack between the two elections of 1974 and at the October election both Labour and Conservative manifestos unequivocally committed their parties to devolution: 'The next Labour Government will create elected Assemblies for Scotland and Wales' (Labour Manifesto, October 1974); 'In Scotland we will set up a Scottish Assembly' (Conservative Manifesto, October 1974). The Liberals, the SNP, the Communists and the Greens also backed a Scottish assembly at that election. It is one of the minor paradoxes of British politics that after such universal commitment no assembly was set up in the 1970s! But the similarity of Labour and Conservative commitments did not reflect a bipartisan policy. Even when they had the same policy it was an issue between them. Briefly in 1974–75 it was, in Butler and Stokes' terms a 'valence issue', before it reverted to being a classic 'position issue'.[23] Labour and the Conservatives only briefly occupied the same ground as they swept past each other on the battlefield.

When Thatcher succeeded Heath as Conservative leader she told a rally in Glasgow that a devolved Scottish 'Assembly must be a top priority to ensure that more decisions affecting Scotland are taken in Scotland by Scotsmen [sic]' – but within a year she switched policy and became totally opposed to devolution. Labour still had some reservations, but its battle with

Thatcher's Conservatives on the issue helped to convince it that its new policy must be correct. By instinct, Labour knew that nothing which irritated Thatcher's Conservatives so much could be bad. The parties polarised on the issue – in their new positions of course. During the 1980s Labour became increasingly committed to devolution, even tinged with a degree of nationalism and the Conservatives swung increasing behind the new Thatcherite party line. In 1976, at a key vote on the need for a guillotine to help the devolution bill through parliament, only six of the 16 Scottish Tory MPs obeyed Thatcher's whip and half of those who did retired or were defeated at the next election. Six Conservative spokesmen, including the Shadow Secretary of State for Scotland and the party chairman resigned. George Younger and Rifkind, both later Conservative Secretaries of State for Scotland supported devolution in the 1970s, as did Michael Ancram, appointed in 1997 by William Hague to oversee constitutional issues and lead the attack on devolution. In 1975, the *Scotsman* reported Ancram's claim to be 'a long time supporter of devolution and the setting up of a directly elected assembly' and his view that the 'English backlash' was the 'unconsidered reaction of a group of people who have been asleep to the realities of the devolution argument for far too long'. By 1997 he was saying: 'we are arguing at the moment [sic!] that the status quo is the best of the options on offer.'[24]

DEVOLUTION COULD ONLY BE ACHIEVED IN CONDITIONS THAT MADE IT UNNECESSARY

Ironically Scotland could only achieve devolution in conditions that made it unnecessary. The SNP have been ideologically opposed to devolution at least since the 1940s. There never seemed much prospect of the SNP winning so many seats in Scotland – whether a bare majority or more would suffice is open to debate – that it could claim a credible mandate to renegotiate the constitutional settlement with England. And if it had done so, it would have demanded independence not devolution. Thus, devolution, if not independence, could only be achieved by English votes. After Thatcher turned her party against devolution, that meant it could only be achieved when the English electorate rejected Thatcher or her Conservative followers and elected a Labour government.

In 1974 it could be argued that Labour adopted devolution as a sop to Scottish national sentiment, fearful that Scottish voters might desert en masse to the SNP. By the 1990s, however, the strongest motivation for Labour to support devolution was to find some means to shelter Scotland from the Thatcherite policies of the London government, not to pander to nationalism. It was frequently pointed out that no Labour-controlled Scottish assembly would have imposed the hated poll tax on Scotland. But in so far as Labour was able to shelter Scotland from Thatcherite policies, a Labour victory in Britain would allow it to shelter the whole of Britain from such policies. No Labour-controlled British government would impose anything like the poll tax anywhere in Britain.

Devolution became a real prospect after the 1997 election when the Lord Chancellor,[25] the Prime Minister, the Foreign Secretary and the Chancellor of the Exchequer in the new Labour government were all Scots. (Altogether, nine of the 23 members of the British Cabinet were Scots.) According to a *Scotsman*/ICM poll, Tony Blair was then the most popular politician in Scotland, far more popular than any of the MPs from Scottish constituencies.[26] Why would Labour want to set up a devolved Scottish assembly to shelter Scotland from that? To transfer control of Scottish affairs from Labour's (Scottish-led) first 11 to its second 11? If Labour's commitment to devolution had not already existed, is it likely that it would have invented it in 1997? This absurdity might be a paradox if it could be argued that Scottish devolution might shelter a Labour-controlled Scotland from a future Conservative-controlled British government. To a degree that might happen, but only if the future British Conservative government treated the Scottish assembly as more federal than devolved, whether out of self-denying principle or simply to avoid the legal and political hassle of overruling the Scottish assembly. Those who point to the example of Stormont in this respect, neglect both the unique distaste of British politicians for involvement in Irish affairs, and their quick abolition of Stormont when they deemed it necessary. It is unlikely that a British government would feel quite so inhibited about interfering in British mainland affairs.

SCOTTISH POLITICAL CULTURE IS *NOT* DISTINCTIVE

Scottish culture is visibly different from that in England, but the visible differences are either superficial or irrelevant to politics. A loch full of whisky and a mountain of tartan wrapped shortbread do not constitute a political culture – though they may help to reinforce a sense of identity. Much more surprisingly, even the existence of a distinctive history and separate institutions does not guarantee a distinctive political culture. Centuries of living under a legal system founded upon principles that are radically different from those in England seems to have left little impression on the Scottish public outside the salons of Edinburgh lawyers. The famous egalitarianism of the Scots is grossly exaggerated, and they are at least as authoritarian as people elsewhere in Britain.[27] For all their anti-Conservative votes, Scots are more conservative (with a small 'c') on abortion, for example[28] – a fact that led Blair to reserve abortion policy to Westminster rather than devolve it to the Scottish assembly.[29] Certainly there are some small differences of political culture between people in Scotland and England, but they are trivially small compared to the differences of political culture within both nations – between young and old, between rich and poor, or between those with high and low levels of education, for example.

Throughout the long years of Thatcherism, Scottish votes diverged increasingly from the English. In 1987, at the height of this divergence there was a gap of 35 per cent between the Conservative lead over Labour in Scotland and England, but, writing in the early 1990s David McCrone noted the 'irony that ... Scottish political behaviour has never in post-war politics been so divergent from its southern counterpart, a situation seemingly achieved with little help from "Scottish National Culture". Here it seems, is a political manifestation which is not tied to a specific cultural divergence ... this expression of political difference ... has developed without encumbrance of a heavy cultural baggage.'[30]

Of course, there is a related paradox that political conflict does not presuppose cultural differences. It can be based as much on differences of interest and identification as on differences of political values. The American revolutionaries did not reject British political culture and values; they rejected British domination. In the 1970s disagreement about whether North Sea oil should be considered a Scottish or a British resource implied no difference in political culture, merely a similarity of avarice combined with a similar, but opposed, sense of Scottish/English identification. In consequence, driven by a separate sense of Scottish identity, by English antagonism, or even by misplaced English generosity, Scotland could win its independence yet use that independence to construct a clone of the English polity in northern Britain – separate but with the same political culture. There is little evidence that any large number of Scots want greater independence for any purpose other than greater independence itself. Divergent voting trends since 1979 grossly exaggerate the policy differences between the Scots and the English. Even while reporting massive popular support for a Scottish assembly in 1997, the Scottish press was already complaining that the new Labour was *not* imposing the same educational reforms on Scotland as on England.[31]

DEVOLUTION IS *NOT* AN IMPORTANT ISSUE FOR THE SCOTTISH
PUBLIC

If voting differences between Scotland and England do not reflect policy
differences on the standard issues of politics, what do they reflect? Perhaps
the devolution issue, the issue of Scottish government itself? Unfortunately
the issue has never ranked high in surveys of Scottish opinion. Typically, a
week or two before the 1997 election, devolution/independence ranked
seventh in the Scottish public's list of concerns.[32]

This has long been an intriguing paradox. So much of what happens in
Scottish politics seems explicable in terms of the debate over devolution:
the Conservatives' relative advantage (once class differences were taken
into account) in the 1950s when they were the more nationalist of the big
two parties, the increasing Labour advantage which can be traced to the
point in 1975 when Thatcher switched from supporting devolution to
opposing it,[33] and many of the short-term trends and the blips in support for
the parties. Similarly, at any one time there is a close correlation, which has
strengthened over the years, between support for independence, devolution
or the status quo on the one hand, and support for the SNP, Labour or the
Conservatives on the other.[34]

Yet when asked about the most important issues in an election campaign,
devolution ranks near the bottom of the list of significant concerns. And
evidence from surveys, particularly in the weeks before the 1979
referendum, suggests that the strong and growing correlation between party
choice and devolution attitudes owes much to party supporters falling into
line with their party's official policy on the issue rather than defecting to the
party that best represents their attitudes towards independence or
devolution.

Definitive evidence is elusive, but party policy positions on devolution,
while not overwhelmingly important in themselves, may be a means of
identifying the party as 'Scottish' or 'foreign'. Surveys showed that by the
1990s a large majority of Scots had come to regard the Conservative Party
as an essentially 'English' and 'foreign' party, out of touch and out of
sympathy with Scottish people.[35] Books by Ian Lang and others asserting the
Conservatives' claim to be the oldest party in Scotland, pictures of Michael
Forsyth in kilts, the solemn repatriation of the Stone of Destiny to
Edinburgh after 700 years in England – none of these seemed to make the
slightest difference to the perception of Conservative 'foreignness'. The
problem was not so much that they were mere symbols, but that they were
merely historical symbols, classified like *Braveheart* under 'entertainment,
sport and tourism' rather than, as such historical symbols would so

tragically be in Ireland, under 'current politics'. Though we cannot be sure, it seems likely that though devolution may not have been directly important *as an issue*, it was important *as a symbol* and, unlike the Stone of Destiny, a symbol with contemporary and political significance.

PEOPLE IN SCOTLAND DO *NOT* IDENTIFY THEMSELVES AS SCOTTISH

'Cars have lights at the front' – that is not wrong, but could be misleading, for they also have lights elsewhere. So it is with national identity in Scotland: it is not wrong, but it is none the less very misleading to say that Scots have a sense of Scottish national identity. Indeed, it is more misleading than informative to suggest, even by default, that the national identity of Scots is encompassed by their Scottishness. The part is not the whole. To that extent it is a paradox rather than an absurdity to claim that the people of Scotland 'do *not* identify themselves as Scottish': unlike its antithesis (that the people of Scotland 'do identify themselves as Scottish') it is manifestly absurd and yet it contains just as much truth – no more, but no less.

The SNP's parliamentary breakthrough in 1974 drew attention to what was distinctive about Scotland – and that clearly included a sense of Scottish identity as well as a whole range of specifically Scottish institutions. Some explanations of rising support for the SNP attributed it to a rising sense of national consciousness, increasing identification with Scotland.[36] Certainly the vast majority of these who lived in Scotland claimed to be Scots, whatever they might mean by that. Other explanations of SNP success stressed the interaction between regional economic differences and Scottish identity, or between the North Sea oil bonanza and Scottish identity. But some focused on a second identity within Scotland: British identity. Declining identification with the British empire or with British national politics, could at least facilitate the growth of Scottish (or Welsh) identity and support for separatist parties. When survey researchers asked people in Scotland to choose between a Scottish or a British identity their choice did prove a useful predictor of political behaviour, especially electoral behaviour.

But there was always something unreal about that choice. At best it was a forced choice. It did not divide those who identified slightly more with Scotland than with Britain (or vice versa) from those who identified exclusively with one or the other. We need to distinguish between vertical and horizontal, hierarchical and non-hierarchical identities. Most people in Britain would have no problems choosing between a Scottish or English identity, or between a British or French identity. These are horizontal, non-hierarchical, basically exclusive alternatives – a few immigrants or children of mixed marriages apart. But people do have problems choosing between a British and a Scottish identity. To the dismay of the minority of nationalist zealots, these are *not* exclusive alternatives for most people. If forced, people will choose between them; but when given a free choice, many in Scotland claim to be equally British and Scottish – and most claim at least to be 'both British and Scottish' if not equally so. In 1992 a third of Scots claimed to be 'equally British and Scottish', and four-fifths to be 'both British and Scottish' in some measure.[37] As the balance of national identity ranged from 'only British', through 'more British than Scottish', 'equally British and Scottish', and 'more Scottish than British', to 'only Scottish', support for independence rose monotonically from four per cent to 45 per cent, and SNP votes from nought per cent to 43 per cent.[38]

But to say that someone feels 'more Scottish than British' is not to say *how much* more Scottish than British they feel. We can quantify the balance of identities, and extend the notion of multiple hierarchical identities to more than two levels of territory. In the 1994–95 Local Governance Survey we asked people throughout Britain to tell us on a scale from nought to ten, *how much* they identified (literally: 'felt a sense of belonging to') their local government district, their region (of which 'Scotland' was one, 'Wales' another, along with eight regions of England), Britain, Europe, the places where they were born, where they lived, and where they worked – and a variety of other things including their family, a religion, a class and a political party.[39]

Identification was strongest with 'family' and weakest with 'a political party'. More relevant to the present discussion is the pattern of identification with territory. Identification with 'Britain' was strong, not so strong as with 'family' but slightly stronger than with people's 'circle of friends'. It averaged 8.0 on the ten-point scale. Region within Britain scored 7.7 however, ahead of home neighbourhood at 7.3, local council district at 7.0, work place at 5.5 and Europe at 4.9 – which was nonetheless still slightly ahead of 'a religion' or 'a political party'. Within different regions of Britain, the balance between British and regional identification varied. In the English South and Midlands, identification with Britain ran well ahead of identification with the region; in Yorkshire and the northern England, identification with the region ran about equal to identification with Britain; and in Scotland and Wales (but more especially in Scotland) identification with the region ('Scotland' or 'Wales') ran well ahead of identification with Britain.

But throughout Britain, territorial identification was clearly multiple: it was simply not the case that those who identified most strongly with their region were the most likely to reject an identification with Britain. That would imply a negative correlation between British and regional identifications, which did not happen. Indeed, quite the opposite: in the sample as a whole there was a substantial positive correlation of 0.31 between identification with Britain and with the region. Certainly that correlation varied: it ran at 0.41 in southern England, 0.44 in northern England, and 0.45 in Wales, though only 0.10 in Scotland. Nonetheless, it remained positive, not negative, even in Scotland. Suppose we categorise those who ranked at least six out of ten, as indicating a strong identification (Table 2). By that criterion, a minority of 20 per cent in southern England identified strongly with Britain but not their region; and minorities of 20 per cent in Wales and 32 percent in Scotland identified strongly with Wales or Scotland but not with Britain; but a large majority, at least 60 per cent in every part of Britain including Scotland, identified strongly with both Britain and their region – and the exact balance between these twin positive identities is less significant than the fact that people identified, not just weakly but strongly, with both.

TABLE 2
PERCENTAGE OF THOSE WHO IDENTIFY POSITIVELY WITH BRITAIN AND REGION
(RANK SIX OR MORE OUT OF TEN)

	Neither	Region only	Britain only	Both
In South and Midlands of England	7	5	20	68
In North of England including Yorkshire	4	7	10	79
In Wales	7	20	10	64
In Scotland	6	32	3	60

Even that insight distorts the reality, however, which is that people switch quickly, easily and unselfconsciously between emphasising different components of their complex national identity according to context and circumstances: at a Scotland-England rugby match they may focus on the Scottish component, yet switch their emphasis to the British component when faced with an Argentinian invasion of the Falklands – or with proposals to set up a separate Scottish state complete with its own army, navy and air force. To put it another way: hierarchical identities are typically both multiple and contingent.

CONCLUSION

I have discussed a long list of potential paradoxes in detail, alluded more briefly to others, and even footnoted at least another two. Their status as paradoxes rather than mere absurdities depends upon the perspective of the reader as well as the self-contradictions of politics and politicians themselves. On some, I have set out my own view of their status, but I have not attempted a systematic classification. It is unlikely that there will be a consensus on which are paradoxes and which mere absurdities. But I hope no one will dismiss *all* of these potential paradoxes as mere absurdities. I hope I have established that there is plenty of paradox in the politics of periphery – though the paradox is as much *about* the periphery as it is *on* the periphery. Politicians and commentators at the centre are guilty of at least as much muddled thinking and emotional prejudice about the West Lothian question, subsidiarity, sovereignty, or devolution as any on the periphery itself.

NOTES

1. Roughly summarised, the Belgrano question was this: was the Argentine cruiser *Belgrano* steaming towards or away from the Falkands in 1982 when Thatcher gave the order to sink her with great loss of life? See Tam Dalyell, *Misrule: How Mrs Thatcher has Misled Parliament from the Sinking of the Belgrano to the Wright Affair* (London: Hamish Hamilton 1987).
2. Lawrence Donegan, 'Who will speak for those who say "No"?', *Guardian* 19 June 1997, p.19.
3. Quoted and debated by Lord Jenkins, 'The case for a People's Bill of Rights', in W.L. Miller (ed.) *Alternatives to Freedom: Arguments and Opinions* (London: Longman 1995). See also Albert Venn Dicey, *An Introduction to the Study of the Law of the Constitution*, 10th ed. (London: Macmillan 1959; orig. 1885).
4. MP for Cathcart in Scotland and leader of the Scottish Conservatives until defeated at the 1979 general election; elected MP for the Essex constituency of Southend East at the 1980 by-election.
5. See e.g. W.L. Miller, A.M. Timpson and M. Lessnoff, *Political Culture in Contemporary Britain: People and Politicians, Principles and Practice* (Oxford: OUP 1996) p.190; or a decade earlier, William L. Miller, Harold D. Clarke, Martin Harrop, Lawrence LeDuc and Paul F. Whiteley, *How Voters Change: The 1987 British General Election Campaign in Perspective* (Oxford: OUP 1980) p.285.
6. See, e.g., W.L. Miller, 'Politics in the Scottish City 1832–1982' in George Gordon (ed.) *Perspectives of the Scottish City* (Aberdeen UP 1985).
7. E.g.: 'Labour Betrays its Promises', *Scotsman* 27 June 1996; 'Faithful Sold Out as Sights Set on South', *Scotsman* 28 June 1996.
8. See Bill Miller, 'Tax factor and devolution', *Glasgow Herald* 28 Dec. 1991, p.9. The analysis was based on 14,216 interviews in BSAS surveys for 1983, 1984, 1985, 1986, 1987 and 1989.
9. As Butler and Pinto-Duschinsky note: 'The Boundary Commission's proposals meant, according to various estimates, a gain of between five and 20 seats for the Conservatives ... [so] the House of Commons went through the charade of Mr Callaghan [the Labour Home

Secretary at the time] presenting the Commissioners' proposals and asking his party to vote them down'. David Butler and Michael Pinto-Duschinsky, *The British General Election of 1970* (London: Macmillan 1971) p.46. Labour lost the 1970 general election despite this blatant gerrymandering, and the new Conservative government promptly implemented the Commissioners' proposals.

10. See Vernon Bogdanor and David Butler (eds.) *Democracy and Elections* (Cambridge: CUP 1983).

11. See G. Bingham Powell, *Contemporary Democracies: Participation, Stability and Violence* (London: Harvard UP 1982) esp. Ch.10 for an impressive attempt at drawing up the balance-sheet of practical virtues and vices associated with PR, FPTP and presidential electoral systems.

12. In its pure form, the extreme option of a presidential system would create more problems than it solved by conferring a frightening excess of power on a unified executive. So, in practice, presidential constitutions tend to be anti-presidential in their details. See 'Paradoxes of Presidentialism' in Juan J. Linz, 'The Perils of Presidentialism', originally published in the *Journal of Democracy* 1/1 (Jan. 1990) pp.51–69 and reprinted in Arend Lijphart (ed.) *Parliamentary Versus Presidential Government* (Oxford: OUP 1994) pp.118–27.

13. Labour Party, *New Labour – Because Britain Deserves Better* (London: Labour Party 1997) pp.33–4.

14. Advocated earlier in Jim Sillars, *Scotland: the Case for Optimism* (Edinburgh: Polygon 1986) Ch.10: 'Scotland within Europe: the Framework for Independence'.

15. John Peterson 'Subsidiarity: a Definition to Suit any Vision?', *Parliamentary Affairs* 47/1 (Jan. 1994) pp.116–32 at p.117.

16. For a contemporary perspective on the plight of the unfortunate Pope Pius XI see Herman Finer, *Mussolini's Italy* (London: Gollancz 1935; repr. NY: Grosset and Dunlap 1965) Ch.16: 'Social Policy'. In an encyclical discussed by Finer, the Pope denounced 'a real pagan worship of the State which is in no less contrast with the natural rights of the family than it is in contradiction with the supernatural rights of the Church' (p.465). The Fascist State had, for example, dissolved all Catholic youth organisations in 1927 because they competed with its own youth organisation, the Balilla.

17. See Andrew Scott, John Peterson and David Millar, 'Subsidiarity: a 'Europe of the Regions' v. the British Constitution?', *Journal of Common Market Studies* 32/1 (March 1994) pp.47–67.

18. *Northern Ireland House of Commons Debates* Vol.16, Col.1095, 24 April 1934.

19. Dicey championed 'the absolute legislative sovereignty or despotism of Crown in Parliament' in his *Introduction* (note 3), but Bogdanor points out that 'it is paradoxical that Dicey should have been the first to advocate the referendum in Britain' when parliament displeased him over Irish home rule. Vernon Bogdanor, 'Western Europe', in David Butler and Vernon Bogdanor (eds.) *Referendums around the World* (London: Macmillan 1994) p.34. For the original see Albert Venn Dicey, 'Ought the Referendum to Be Introduced into England [*sic*]' *Contemporary Review* (1890) pp. 505, 507; or Albert Venn Dicey, *A Leap in the Dark* 2nd ed. (London: John Murray 1911).

20. Lord Hailsham, *The Dilemma of Democracy: Diagnosis and Prescription* (London: Collins 1978) Ch.20: 'Elective Dictatorship'. Unionists and Conservatives only seem to believe in parliamentary sovereignty when their party controls Parliament.

21. See W.L. Miller, *The End of British Politics?* (Oxford: OUP 1981) pp.21–2, based on contemporary reports in the Scottish press.

22. Ibid. p.49.

23. See David Butler and Donald Stokes, *Political Change in Britain: The Evolution of Electoral Choice*, 2nd ed. (London: Macmillan 1974) p.292.

24. Ancram's changing position is set out clearly in the *Scotsman*, 23 June 1997, p.4 from which both quotations are taken.

25. The chief law officer and highest-paid member of the British cabinet.

26. For these ICM poll results see the *Scotsman* 19 June 1997, p.8. The same poll also showed

that the Liberal leader, Paddy Ashdown, MP for a constituency in the south-west of England, was the second most popular politician in Scotland – ahead of Donald Dewar Labour's Scottish leader, Jim Wallace the Liberals' Scottish leader, and Alex Salmond leader of the SNP. John Major and the Conservatives' Scottish chair, Annabel Goldie, came far behind with negative ratings.

27. See Miller *et al.* (note 5) pp.364–73.
28. Ibid. p.151.
29. See 'Labour at War over Scots: Cabinet Split on Devolution', *Guardian*, 8 July 1997, p.1; 'Pro-Life Groups Attack Decision over Abortion Law', *Scotsman*, 9 July 1997, p.4. Blair feared the growth of a trade in cross-border abortions between Scotland and England, as already existed between Ireland and England.
30. David McCrone, *Understanding Scotland: The Sociology of a Stateless Nation* (London: Routledge 1992) pp.195–6.
31. 'Scotland's Schools Miss out on Reform', *Scotsman*, 8 July 1997 p.1 – the lead story.
32. ICM poll, *Scotsman*, 17 March 1997, p.4. Devolution or a Scottish assembly came behind the economic situation, unemployment, taxes, welfare, education and the NHS.
33. See W.L. Miller, *The End of British Politics?* (Oxford: OUP 1981) p.232.
34. Compare for example ibid. p.114 with Alice Brown, David McCrone and Lindsay Paterson *Politics and Society in Scotland* (London: Macmillan 1996). The comparison shows that between the elections of October 1974 and 1992 the percentage of Conservative voters supporting any degree of constitutional change (devolution or independence) dropped from 55 per cent to 41 per cent; the percentage of SNP voters supporting outright independence rose from 48 per cent to 57 per cent; and the percentage of Labour voters supporting devolution (but not independence) rose from 42 per cent to 62 per cent – all representing increasing harmony between voters' attitudes and party policies: An ICM poll in 1997 showed little further after 1992; see *Scotsman*, 3 April 1997, p.13.
35. An ICM poll showed 73 per cent of Scots agreed with the statement: 'The Conservative Party is a mainly English party with little relevance to Scotland': *Scotsman*, 17 Feb. 1997, p.5. Compare the Gallup Poll finding that 31 per cent of Scots chose 'The Conservative Party is a mainly English party which governs Scotland in England's interests' as the main reason for the Scottish Conservatives' decline, while only 8 per cent chose 'Conservatives oppose giving Scots any form of self-government'. There were other reasons for the Scottish Conservatives' decline, of course. See *Daily Telegraph*, 9 May 1996, p.11.
36. Jack Brand, *The National Movement in Scotland* (London: Routledge 1978).
37. Lynn Bennie, Jack Brand and James Mitchell, *How Scotland Votes* (Manchester: Manchester UP 1997) p.133.
38. Ibid. p.140.
39. For more details of this survey and a further discussion of these measures of multiple identities, as applied to the general public, elected local councillors, and appointed members of quango boards, see Bill Miller and Malcolm Dickson, 'The Democratic Principle in Local Governance', in Gerry Stoker (ed.) *Power and Participation: the New Politics of Local Governance* (London: Macmillan 1998, forthcoming).

The British Electorate in the 1990s

DAVID DENVER

During the 1990s the British electorate has displayed a remarkable capacity to surprise and confound media commentators, political pollsters and academic psephologists. Before the 1992 general election it was widely expected that the result would be close and that the Conservatives would, at best, lose their overall majority in the House of Commons. In the event, they retained their majority and, in terms of share of the vote, won the election by a wide margin (see Table 1, p.198). In 1997, however, there was a remarkable turnaround. In this case it was not that Labour's victory was a surprise, but the scale of the victory. Although the national opinion polls consistently predicted a Labour landslide during the election campaign, most commentators simply did not believe that such a landslide was possible and discounted the poll results. As Table 1 shows, however, an almost unprecedented electoral tidal wave swept Labour to power and also returned 46 Liberal Democrats to the House of Commons. Although they took many people by surprise, it cannot be said that there is anything intrinsically paradoxical about the results of these two elections. When they are placed in the context of recent studies of voting behaviour in Britain, however, there clearly are paradoxical elements.

The major report on the 1992 election by the British Election Study team was entitled *Labour's Last Chance?* reflecting a commonly held view that very serious obstacles stood in the way of Labour's ever again winning an election. Although the authors argued that changes in the social structure did not condemn the party to inevitable defeat in future, these changes meant that Labour was 'trying to sail against the current' in its efforts to win.[1] In similar vein, Richard Rose, surveying election results and social trends since 1964, suggested that Labour was in structural decline.[2] Yet five years later in 1997 – with social changes having continued to favour the Conservatives in the interim – Labour scored an overwhelming victory. Moreover, the incumbent Conservatives were crushed despite the economy being relatively buoyant, contrary to the received wisdom about the positive relationship between economic performance and the popularity of governments in Britain and elsewhere.[3] Before examining the 1997 election in more detail, however, it is worth considering the four Conservative

TABLE 1
GENERAL ELECTION RESULTS 1992 AND 1997

	1992 Votes		1997 Votes	
	Share (%)	Seats	Share (%)	Seats
Conservative	42.8	336	31.4	165
Labour	35.2	271	44.4	419
Liberal Democrat	18.3	20	17.2	46
Other	3.7	24	7.0	29

Note: Figures for vote share refer to Great Britain only while those for seats refer to the United
       Kingdom as a whole.

election victories from 1979 to 1992. The relative stability of Conservative
support in this period is also somewhat paradoxical since, according to the
dominant view in British voting studies, it had been preceded by a 'decade
of dealignment'.[4]

THE DEALIGNMENT THESIS

There is a broad (but not universal) consensus among students of voting
behaviour that the modern British electorate can be characterised as
'dealigned'. Dealignment refers to two related, but conceptually distinct,
processes. Early voting studies, in the 1950s and 1960s had suggested that
party support closely reflected major social cleavages. There was a clear
alignment between the social groups to which voters belonged (especially
their social class) and the party they voted for. From the 1970s, however,
although new cleavages such as race emerged and regional differences
became more important, the link between social characteristics and party
choice became steadily weaker. In particular, on most measures, class
voting declined. The Alford index of class voting,[5] for example, was 42 in
1964 and 43 in 1966. For the four elections of the 1970s the average index
score was 32; in both 1983 and 1987 it was 25 and in 1992 the score was
27. Most commentators take these figures as evidence of a 'class
dealignment'.[6]

The second aspect of dealignment relates to party identification – the
enduring psychological attachment to a political party, the sense of being a
supporter of a particular party, which was formerly characteristic of many
voters. In 1964, 81 per cent of the electorate identified with either the
Conservatives or Labour and 40 per cent described themselves as 'very
strongly' Conservative or Labour. By 1992, generalised attachment to the
major parties was still common – 78 per cent identified with one of them –

but the strength of attachment had declined significantly. Only 18 per cent were now 'very strong' major-party identifiers. This indicates a clear weakening of the psychological alignment between voters and the major parties – a 'partisan dealignment'.

Taken together, the processes of class and partisan dealignment are the most striking developments in British electoral behaviour during the last 30 years. In particular, the decline in the average strength of party identification among electors – about which there is little or no dispute – would be expected to have had important electoral consequences. Strong party identification acts like an anchor, binding the voter securely to his or her party and thus ensuring stability in party support over lengthy periods (and for a lifetime in many cases). As with those whose vote is determined by their class identity, party choice at elections is close to being automatic for very strong party identifiers. When the anchor is loosened, however, the voter is more likely to drift, to be more affected by short-term forces. A dealigned electorate is likely to be more open to persuasion by the parties and the media, more indecisive about which party to vote for, and more likely to switch parties. Rather than party choice being stable and predictable – as with the aligned electorate of the 1950s and 1960s – we should expect the dealigned electorate of the 1980s and 1990s to be volatile and unpredictable. Paradoxically, however, general elections from 1979 to 1992 displayed striking stability in one important respect: put simply, the Conservatives kept on winning.

1979–92: A VOLATILE ELECTORATE PRODUCING STABLE ELECTION OUTCOMES

The picture that voting studies have drawn of the modern electorate – hesitant, volatile and unpredictable – appears to be contradicted by the Conservative Party's steady performance in the four general elections from 1979 to 1992. They won all four and their share of the vote in Great Britain varied only slightly, from 44.9 per cent in 1979 to 43.5 per cent in 1983, 43.3 per cent in 1987 and 42.8 per cent in 1992. The vote shares gained by Labour and the Liberals and their successors varied much more, of course. It is also the case that in the periods between elections the Conservatives experienced lengthy periods of unpopularity and suffered spectacular losses in parliamentary by-elections and local elections. But when a general election came round voters appeared to forget their mid-term misgivings and regularly returned the Conservatives to power. How is this paradox to be explained?

TABLE 2
PARTY CHOICE OF SELECTED SOCIAL GROUPS 1992

	Conservative %	Labour %	Liberal Democrat %	
All	46	34	17	(2,423)
Non-manual workers	54	23	20	(1,299)
Manual workers	34	49	13	(1,046)
Salariat	54	19	24	(673)
Routine non-manual workers	52	29	17	(574)
Petty bourgeoisie	64	16	18	(167)
Foremen/technicians	39	44	15	(119)
Working class	31	54	12	(813)
Owner occupiers	53	27	18	(1,800)
Council tenants	18	65	13	(429)
Aged 18-34	43	36	18	(696)
Aged 35-54	46	32	18	(869)
Aged 56-64	44	36	17	(362)
Aged 65+	49	36	14	(471)
Men	44	36	17	(1,123)
Women	47	33	18	(1,301)
Trade union members	30	48	19	(490)
Non-trade union members	50	33	18	(1,301)
No religion	40	38	19	(727)
Church of England	54	28	18	(1,006)
Church of Scotland	39	30	10	(121)
Roman Catholic	33	50	14	(245)
Other Protestant	46	35	17	(181)
Other	43	37	17	(136)

Source: British Election Study 1992 cross-section survey.
Note: Rows do not total 100 because votes for other parties are not shown. Figures in brackets
         indicate the number of cases on which the percentages are calculated.

The first point to note is that the social changes of the past 30 years have advantaged the Conservatives. Although the association between the social characteristics of voters and party choice has weakened, it has not entirely disappeared. This is illustrated in Table 2 which shows the party choice of selected social groups in 1992. As can be seen, class differences persisted, whether the broad manual/non-manual division or the more elaborate five-class scheme proposed by Heath *et al.*[7] is used to categorise respondents. There were also sharp differences between owner occupiers and council tenants. On the other hand, differences in terms of age and sex were relatively small. Trade union members were more likely to vote Labour than non-members and party choice also continued to be related to religious affiliation.

The existence of these sorts of variations in voting behaviour has been, of course, documented in successive voting studies. What is important in this context, however, is that most of the groups which have traditionally favoured the Conservatives have been increasing in size, while those that favoured Labour have been declining. The most notable change in this respect is the growth of the middle class. At the time of the 1966 census 11.5 per cent of the British workforce had professional or managerial occupations and the broader non-manual grouping comprised 44 per cent. By 1991 the proportion of professional and managerial workers had grown to 33 per cent and that of the non-manual class to 56 per cent. Also by 1991, 66.5 per cent of households were owner occupiers compared with only 45 per cent in 1966. To the extent that class and housing tenure still influenced party choice in the early 1990s, then, the pool of 'naturally' Conservative voters had grown considerably since the 1960s. Other changes favouring the Conservatives included the steep decline of heavy industries, such as coal-mining, steel-making and shipbuilding; the growth of service employment; a slow but steady movement of population from Scotland and the North to the South and from cities to suburbs and rural areas; a sharp decline in trade union membership; and a major extension of share ownership from a very small minority to around 20 per cent of the adult population. There were trends working in the opposite direction – a growth in the numbers of ethnic minority voters and of working women, for example – but it has been calculated that the Conservatives' share of the vote in 1992 was four percentage points greater (and Labour's share five points smaller) simply because of changes in the sizes of the electorally most important social groups from 1964.[8]

None the less, changes in the social structure alone cannot account for the string of Conservative election victories from 1979 to 1992 and dealignment theorists have tended to play down their importance. Their favoured explanation focuses on electors' opinions about issues. In this view, it is a mistake to regard the 40-odd per cent of the vote that the Conservatives consistently garnered from 1979 to 1992 as constituting a solid bloc of core supporters. Data on party identification from the relevant British Election Study surveys show that the true Conservative core – those who identified very strongly with the party – never amounted to more than nine per cent of the electorate in this period. The interpretation which is consistent with the picture of a dealigned electorate is that, at successive

TABLE 3
ISSUES IN ELECTIONS 1979–92

| | 1979 | | 1983 | |
	Salience	Preferred party lead	Salience	Preferred party lead
Prices	42	Lab +13	20	Con +40
Unemployment	27	Lab +15	72	Lab +16
Taxes	21	Con +61	–	
Trade unions/strikes	20	Con +15	–	
Law and order	11	Con +27	–	
Defence	–		38	Con +54
NHS	–		38	Lab +46

| | 1987 | | 1992 | |
	Salience	Preferred party lead	Salience	Preferred party lead
Unemployment	49	Lab +34	36	Lab +26
Defence	35	Con +63	3	Con +86
NHS	33	Lab +49	41	Lab +34
Education	19	Lab +15	23	Lab +23
Prices	–		11	Con +59
Taxes	–		10	Con +72

Sources: Reproduced from a table in D. Denver, 'Electoral Support' in P. Norton (ed.) *The Conservative Party* (Hemel Hempstead: Prentice Hall 1996). The original Gallup data were analysed by Ivor Crewe in a series of post-election reports.

Note: The table reports issues mentioned by more than 10 per cent of survey respondents at any election. The preferred party lead is the percentage saying the party most preferred on the issue had the best policy minus the percentage saying the same of the next most preferred party.

elections, the Conservatives were able to put together a temporary coalition of voters, which then dissolved in the inter-election periods. These coalitions were based on electors' short-term judgements concerning political issues that were prominent during election campaigns.

Evidence relating to issue voting in elections from 1979 to 1992 is given in Table 3. This shows the salience of different issues – the percentage of respondents mentioning an issue as one that they thought important in deciding how they would vote – and the party preferred by the electorate on the issue concerned. The argument of dealignment theorists is that a combination of issue salience and judgements about which party had the best policies provides a satisfactory explanation for the Conservative victories in these elections. The data, derived from Gallup's election polls, suggest that the argument has some force for 1979 and 1983. In 1979, although Labour was preferred on the most salient issues (prices and unemployment), their lead was relatively small. The Conservatives, on the other hand, enjoyed a huge lead on taxation and were significantly ahead on the other two salient issues. In 1983, in the wake of the Falklands War and with Labour apparently leaning towards unilateral nuclear disarmament, the defence issue provided a major electoral bonus for the Conservatives. Along with their large lead on prices, this was enough to outweigh Labour's popularity on the National Health Service and their lead on unemployment. In 1987 and 1992, however, the data suggest a problem for the issue voting hypothesis. The electorate clearly preferred Labour's policies on salient campaign issues but proceeded to return the Conservatives to power. In 1987 the Conservatives retained their position as the party most favoured on defence but Labour was comfortably ahead on each of the other three salient issues. Again in 1992 Labour led on the three most salient issues and, although the Conservatives were overwhelmingly preferred on defence, this issue had dropped out of the election agenda being mentioned as important by only three per cent of Gallup's respondents. If voting had been based entirely on the electorate's assessment of the issues, Labour would have won both elections.

This apparent disjunction between election results and what the electorate thought about the parties' policies on campaign issues in 1987 and 1992 was echoed in other data on electors' policy preferences. A major theme of Conservative campaigning in this period was the need to reduce the burden of taxation and to squeeze public expenditure in order to do so. To explore electors' attitudes on this question, the BES asked respondents to their 1983, 1987 and 1992 election surveys to indicate where they stood on a scale running from 'put up taxes a lot and spend much more on health and social services' to 'cut taxes a lot and spend much less on health and social services'. The results are summarised in Table 4.

TABLE 4
ELECTOR'S OPINIONS ON TAX CUTS AND WELFARE SPENDING 1983-92

	1983 %	1987 %	1992 %
Put up taxes and spend more	43	61	67
Undecided	36	26	21
Cut taxes and spend less	21	13	12
	(3,949)	(3,698)	(1,379)

Source: BES cross section surveys 1983, 1987 and 1992.
Note: Figures in brackets indicate the number of cases on which the percentages are calculated.

In 1983, 36 per cent of respondents placed themselves at the mid-point of the scale, indicating indecision, but those favouring increased taxes outnumbered those who preferred tax cuts by two to one. In the two later elections the gap widened much further with only small minorities taking the position broadly advocated by the Conservatives. Similar figures were regularly reported by opinion polls[9] and these findings were instrumental in persuading Labour strategists in 1992 that there would be no electoral danger in proposing modest increases in taxation. This seems to have been a mistake. When it came to polling day 49 per cent of respondents to the Harris exit poll said that they would be worse off under Labour's tax and benefit policies while only 30 per cent said that they would be better off. Only 7 per cent of the former voted Labour compared with 82 per cent of the latter.[10] It would appear that voters were willing, in general, to subscribe to the view that taxes should be raised (or at least to tell pollsters that they favoured this view) but, paradoxically, when it came to the crunch they voted against the party that proposed to raise their taxes.

The paradox of stable Conservative support among a dealigned electorate cannot, then, be satisfactorily explained by reference to issue voting – at least on an elementary conception of what issue voting involves. Rather, looking at the public's opinions on issues opens up a new set of problems in that, while electors seemed to prefer Labour on important policy areas, they continued to vote for the Conservatives.

In an attempt to deal with these problems, election analysts have gone beyond exploring electors' preferences and perceptions of party stances to include also their estimation of the credibility or competence of the parties on the issues concerned. Here the Conservatives had a decisive advantage. Most commentators agree that the state of the economy is the central issue in determining whether or not voters support the governing party. The economy is not a 'position' issue on which people take sides, such as the privatisation of publicly owned industries. Everyone wants economic prosperity and the issue is how well the parties have performed, or might perform in future, in delivering it. In general, during the lengthy period of Conservative domination, the electorate adjudged them to be the party which was more competent at handling the economy. In the 12 months before the 1992 election, for example, when asked by Gallup which party was best able to cope with Britain's economic difficulties, an average of 45 per cent of respondents opted for the Conservatives and 30 per cent for Labour, despite the fact that the Conservatives had been in office for 12 years and were presiding over yet another economic recession. Similarly, during the 1992 campaign 42 per cent believed that Britain as a whole would be worse off under Labour, and 44 per cent that they and their families would be worse off, compared with 38 per cent and 30 per cent respectively who thought that things would be better. In all four elections from 1979 the Conservatives were normally the preferred party on economic issues, with the exception of unemployment. As we have seen, Labour had large leads on the issue of unemployment, but voters were sceptical about whether any party could do anything about it. As Newton and others have argued,[11] it seems that the electoral impact of high level unemployment declined very substantially during the 1980s. Previously unemployment had been seen as a touchstone of government economic performance and it was thought that any government which allowed it to rise significantly would pay dearly at the polls. In the 1980s, however, voters appeared willing to tolerate – or did not blame the government for – levels of unemployment unheard of since the 1930s.

The question of the effect of economic performance on the pattern of party popularity between elections has received a good deal of attention from academic specialists,[12] although the fact that there is a link between the two has certainly not escaped the attention of politicians. In recent years the most detailed work on this question has been carried out by David Sanders.[13] He found that between 1983 and 1992 the level of support for the Conservatives could be effectively predicted by the aggregate level of personal economic optimism – how people viewed the financial prospects of their household. In turn, economic optimism was influenced by the performance of the 'real' economy, in particular changes in the inflation rate and in interest rates (but not unemployment). Using his model, Sanders was able to make a good prediction of the outcome of the 1992 election 18 months before it was held.[14] On the basis of 'economic voting', explaining the series of Conservative victories after 1979 is not difficult. Once in government, the Conservatives were able to manage the economy in such a way as to ensure that when an election came round personal economic optimism outweighed pessimism among the voters, even if only temporarily, and thus reaped the expected electoral reward.

A further short-term factor helping to explain Conservative successes is the relative popularity of the party leaders. The traditional view, when aligned voting was the rule, was that the competence, character and personality of party leaders counted for little with the electorate. As Ivor Crewe wrote in 1979, 'British voters, if forced to choose between leader and party, tend to abandon the leader.'[15] With the erosion of the influence of long-term factors on party choice, however, party leaders have come to figure more prominently in voters' calculations[16] and at three of the four elections from 1979 to 1992 Conservative leaders have been far more popular than their Labour counterparts (Table 5, p.206). In 1979 the

TABLE 5
BEST PERSON FOR PRIME MINISTER 1979–92

	1979 %	1983 %	1987 %	1992 %
Conservative leader	28	44	40	52
Labour leader	46	13	32	23
Liberal etc. leader	26	37	11	24
SDP leader	–	6	17	–

Source: Gallup election day polls

incumbent Labour prime minister, James Callaghan, was preferred to the Conservatives' Margaret Thatcher but the latter was not particularly unpopular and, when not forced to choose between the two, the electorate rated her quite highly.[17] In 1983, however, Mrs Thatcher had a huge lead over the unfortunate Michael Foot and in 1987 she led Neil Kinnock comfortably. By 1992 John Major had replaced Mrs Thatcher and, with Kinnock still finding it difficult to convince the voters of his qualities, Major was the favoured choice as prime minister of more than half of the voters. It is not possible to attach a precise weight to the influence of party leaders on the electorate but it is hard not to believe that the significantly greater electoral appeal of the Conservative leaders, as compared with their main rivals, goes some way to explaining the failure of an electorate which was Labour-inclined on issues to switch their votes to the Conservatives in significant numbers.

A final point of significance in explaining the Conservatives' electoral hegemony between 1979 and 1992 is that during this period they enjoyed the overwhelming support of the British national press. Up to the mid-1970s the party support of national newspapers was less decisively skewed in favour of the Conservatives than it became during the 1980s and studies of press influence suggested that, in any case, the strength of voters' long-term alignments with their parties was such as to inure them to the blandishments of newspapers. As noted above, however, dealignment theory suggests that when the anchors binding voters to their parties become loosened the voters become more open to persuasion and influence. Moreover, during the 1980s support for the Conservatives and opposition to Labour, became strident to the point of hysteria in the mass-circulation 'Tory tabloids'. Recent research has tended to confirm the expectation of increased press influence. Miller, for example, reported that, in the run-up to the 1987 election, support for the Conservatives increased dramatically among readers of pro-Conservative tabloids, especially among those who were not strongly committed to a party.[18] In their study of press influence between 1987 and

1992, Curtice and Semetko also concluded that there were clear signs of a small but significant longer-term press effect.[19] Evidence that voters are influenced by the newspapers that they read remains somewhat patchy – partly because researching the question is extremely difficult – but the weight of opinion among analysts is that the Conservatives derived some electoral advantage from the strong press backing that they received between 1979 and 1992.

Thus, the paradox of stable Conservative support and consistent Conservative victories, as well as the surprising failure of the electorate's preferences on salient campaign issues to translate into stronger Labour election performances, can be at least partly explained in ways consistent with the thesis of class and partisan dealignment. It remains the case, none the less, that a general increase in electoral volatility failed to materialise to the extent anticipated by dealignment theorists. Two measures of volatility are shown in Table 6. The first, the 'Pedersen index', is derived from election results and is a measure of net volatility.[20] The second, derived from survey data, is the proportion of voters in two successive elections who switched parties, which is the simplest measure of overall (or 'gross') volatility. The peak of net volatility in the elections for which figures are shown was in February 1974, when the Liberals made a dramatic advance, but the 1983 election also saw high volatility – largely because of the impact of the short-lived Social Democrat Party (SDP). In both 1987 and 1992, however, net volatility was lower than in any of the previous elections for which index scores are comparable. The data on party switching suggest that overall volatility increased sharply in the 1970s but stabilised – albeit on a higher plateau – thereafter. In Pippa Norris's words, there appears to have been 'a period-specific shift during the seventies', but no trend to greater volatility in the 1980s and early 1990s.[21]

TABLE 6
ELECTORAL VOLATILITY 1964–92

	Pedersen index	% switchers
1959–64	5.9	18
1966–70	6.0	16
1970–74 (Feb.)	13.3	24
1974 (Oct.) – 1979	8.2	22
1979–83	11.9	23
1983–87	3.2	19
1987–92	5.4	22

Source: A. Heath *et al.*, *Labour's Last Chance?* (Aldershot: Dartmouth 1994) p.281.
Note: Pairs of elections with short gaps between them are omitted.

The concept of electoral volatility involves more than simply vote-switching, however, and referring to other data – on mid-term movements in opinion, the timing of the vote decision and electors' willingness to consider voting for parties other than the one eventually chosen – those who espoused the dealignment thesis argued that the dealigned electorate was at least *potentially* more volatile than in the past. The fact that this potential had not been fully realised did not mean that it did not exist.

1992–97: DEALIGNMENT VINDICATED?

In terms of the trends in party support, the period between the 1992 and 1997 elections was extraordinary. A slump in the popularity of the governing party, followed by a recovery as the next election approaches, is a familiar aspect of the inter-election cycle in Britain (as elsewhere) but the 1992–97 cycle set new records. The Conservatives experienced their worst-ever local election defeats, received their lowest vote share of the century in a national election (27.9 per cent) in the election to the European Parliament in June 1994, and lost all eight of the seats they defended in parliamentary by-elections. In the monthly opinion polls, the Conservative share of voting intentions plunged further and faster than ever before and the gap between the government and the opposition was larger and longer-lasting than in any previous cycle. Between 1987 and 1992, for example, Labour had led the Conservatives by more than ten per cent for 12 months; between 1992 and 1997 Labour's lead was more than 20 per cent for 39 months. On the basis of previous experience, politicians and pundits waited expectantly for the gap to narrow significantly as the 1997 election approached, but it never did. Although the Conservatives recovered a little from their truly disastrous poll ratings in early 1995, they never looked like threatening Labour's lead and were still 25 points behind in March 1997.

During this cycle the Conservatives lost three of the significant electoral advantages that they had previously enjoyed. The first was in terms of leaders. Although John Major had been preferred by the electorate over Neil Kinnock, he soon came to be seen as weak and indecisive. In July 1992 Kinnock was replaced as Labour leader by John Smith. Within three months, Smith overtook Major as the most preferred prime minister, and over his 21-month term as leader he was preferred, on average, by 32 per cent of Gallup's respondents, compared with 22 per cent for Major and 20 per cent for Paddy Ashdown, leader of the Liberal Democrats. Following John Smith's death, Tony Blair became Labour leader in July 1994. Almost as soon as Blair was elected, he established a commanding lead over his rivals as the party leader thought likely to make the best prime minister.

From August to December 1994 he was preferred on average by 41 per cent of the 'Gallup 9000' compared with 16 per cent for Major and 14 per cent for Ashdown. During the next 12 months the average figures were 42 per cent for Blair, 17 per cent for Major and 13 per cent for Ashdown, while for 1996 they were 39 per cent for Blair, 19 per cent for Major and 14 per cent for Ashdown. For the first time for many years the Conservatives faced a general election with a leader who was much less popular with the electorate than his main rival.

The second advantage lost by the Conservatives was their reputation for economic competence. Just five months after winning re-election a central plank of the government's economic policy collapsed when it was forced to withdraw from the European exchange rate mechanism (ERM) in humiliating circumstances. This effectively involved a devaluation of the pound sterling. In terms that all could understand, the media made great play of the fact that several billion pounds had been poured down the drain in a forlorn attempt to shore up the pound, and the voters were left in little doubt that the episode had been a fiasco, that the government had been humiliated and that its handling of the crisis had been hopelessly incompetent. At the time of the 1992 election, when asked which party they thought could handle Britain's economic difficulties best, 49 per cent of Gallup's respondents opted for the Conservatives and only 36 per cent for Labour. In early September 1992 the Conservatives still retained a lead of five points over Labour on this question. In October, however, Labour went into an 18-point lead and this stretched to 21 points in November.[22] Thereafter, the Conservatives never regained their reputation of being the party that could be trusted most to handle the economy competently, although the gap between the parties in this respect narrowed as the election approached.

Largely as a consequence of the ERM fiasco the Conservatives also lost the support of a significant section of the national press. With the Euro-sceptic *Sun* newspaper in the vanguard, the prime minister was portrayed as a weak and indecisive muddler, and he never regained the kind of wholehearted support that Mrs Thatcher had enjoyed. At the start of the 1997 election campaign the *Sun* – the biggest-selling tabloid which had formerly been stridently anti-Labour – ostentatiously announced its support for Tony Blair and called for a Labour victory.

Other aspects of the performance of the Conservatives in government also help to account for the depths of unpopularity which they plumbed. Having been elected on a tax-cutting platform, successive budgets increased taxation. Party divisions over Europe were deep, public and enduring and led eventually to John Major resigning as party leader in June 1995 in order to force a leadership election. He easily defeated his Eurosceptic challenger, John Redwood, but, although this secured Major's position, public squabbling on the issue continued. The Conservatives were also rocked by a series of sexual and financial scandals – no fewer than nine members of the government resigned and a ministerial aide died in the context of a sex scandal – and the government came to seen as tainted by 'sleaze'. In addition, 'boardroom greed' in the newly-privatised public utilities resulted in the public turning decisively against the policy of privatisation. This had been one of the Thatcher governments' most distinctive policies and, although the voters were never very enthusiastic about the privatisations of the 1980s, they were never very hostile either. By the mid-1990s, however, privatisation had become a term of abuse, inextricably associated in the public mind with vastly overpaid executives who sacked workers and increased prices while helping themselves to fortunes. In this context, further privatisations remained a large and highly visible part of the government's programme.[23]

In the inter-election period between 1992 and 1997 the electorate behaved in a way that was characteristically 'dealigned'. As compared with the previous general election high levels of volatility were recorded in local elections, by-elections, European elections and opinion polls. For many voters old loyalties appeared to be forgotten – at least on the part of Conservative supporters – as they reacted to government policies and performance, their assessments of the party leaders and other events. Short-term factors rather than long-term forces determined the pattern of party support and, as a consequence, the Conservatives entered the 1997 election campaign with poorer prospects than any previous post-war government.

The 1997 general election itself was a classic 'dealignment' election in many ways. First, the electorate displayed high levels of volatility. The swing from Conservative to Labour (10.3 per cent) was almost double the previous post-war record (5.3 per cent in 1979) and the vote share of 'other' candidates was close to twice what it had been in 1992. The volatility score on the Pedersen index was 12.5, the second highest since 1945. According to the NOP/BBC exit poll[24] almost 30 per cent of 1992 Conservatives deserted the party and, overall, 23 per cent of those who had voted in 1992 switched parties. Given that these data are based on respondents recalling their 1992 vote, which tends to exaggerate the level of consistency in party support, they suggest high levels of volatility.[25]

Second, although party choice continued to be structured by social characteristics, exit-poll data suggest that class voting reached a historically low level. As Table 7 shows, non-manual workers divided their votes evenly between the Conservatives and Labour. On these data, the Alford index of

TABLE 7
PARTY CHOICE OF SELECTED SOCIAL GROUPS 1997

	Conservative %	Labour %	Liberal Democrat %	
All	31	45	17	(2,332)
Non-manual workers	37	37	20	(1,246)
Manual workers	24	57	13	(844)
Owner occupiers	35	41	17	(1,616)
Council tenants	13	66	15	(254)
Aged 18–29	22	57	17	(372)
Aged 30–44	26	50	17	(667)
Aged 45–64	33	43	18	(737)
Aged 65+	44	34	17	(471)
Men	31	46	17	(986)
Women	32	45	17	(1,250)
Trade-union members	18	57	20	(320)
Non-members of trade unions	34	44	17	(2,012)

Source: NOP/BBC exit poll.
Note: Rows do not total 100 because votes for other parties are not shown. The exit poll contains no information on religion and does not allow analysis in terms of the fivefold class schema presented in Table 2. Figures in brackets indicate the number of cases on which the percentages are calculated.

class voting fell to 20 (the previous low having been 25). Owner occupiers and non-members of trade unions did not depart very sharply from the voting pattern of the electorate as a whole and there was no significant difference between men and women. It is clear, however, that the level of Conservative support increased, and Labour's decreased, with age.

Third, the outcome of the election was clearly consistent with majority opinion among the voters on campaign issues. Table 8 (p.212) shows which party electors thought would handle various issues best, according to Gallup's post-election poll. It is possible, of course, that answers to question like this are in part rationalisations of a decision made to support a particular party on other grounds, but the figures are none the less striking. Labour was preferred on all issues except defence, which hardly figured in the campaign, and even in that case were only narrowly behind the Conservatives. It is not perhaps surprising that Labour should have racked up huge leads on 'natural' Labour issues such as the NHS, education, pensions and unemployment. What is remarkable is that Labour comfortably led on issues which hitherto have been regarded as strong areas for the Conservatives – law and order, taxation, strikes and inflation. Further evidence about the electorate's issue opinions comes from the NOP/BBC exit poll. By 49 per cent to 23 per cent Labour was more trusted than the Conservatives to deal with 'sleaze', by 48 per cent to 26 per cent on education and by 47 per cent to 36 per cent on income tax. Moreover, 78 per cent declared themselves opposed to further privatisation. Clearly the government's record in office had taken its toll and this is also reflected in the fact that 84 per cent of poll respondents thought that the Conservatives were a divided party compared with 41 per cent who thought the same of Labour.

TABLE 8
BEST PARTY ON ISSUES IN THE 1997 ELECTION

	Conservative %	Labour %	Liberal Democrat %	Lead
The National Health Service	14	68	10	Lab +54
Education and schools	15	59	18	Lab +44
Pensions	19	61	7	Lab +42
Unemployment	22	63	5	Lab +41
Law and order	26	53	8	Lab +27
The environment	15	42	23	Lab +27
The unity of the UK	32	46	7	Lab +14
Relations with Europe	30	47	8	Lab +17
Taxation	34	45	10	Lab +11
Strikes and industrial disputes	37	46	6	Lab +9
Inflation and prices	40	43	5	Lab +3
Britain's defence	39	37	7	Con +2

Source: Gallup post-election survey.
Note: Rows do not total 100 because mentions of other parties and 'Don't Knows' are excluded.

Finally, the voters' assessments of the major party leaders were dramatically different from what they had been in the three previous elections. A huge 79 per cent thought that Tony Blair was 'a strong leader', while only 33 per cent ascribed this quality to John Major. In the prime ministerial stakes, Blair was preferred by 47 per cent of exit-poll respondents, Major by 33 per cent and Ashdown by 20 per cent.

All of this evidence suggests that in 1997 the British electorate made its decision on the basis of judgements about party policies, the record of the government in office and the capabilities of the party leaders, rather than falling back upon traditional loyalties. No doubt strong party identifiers and a core vote for each of the parties still exist, but there are now enough voters who can be characterised as 'dealigned' and 'judgmental' to render electoral politics fluid and much more unpredictable than formerly. There remain, however, two paradoxes relating to electoral behaviour in the period up to and including the 1997 election.

The first concerns so-called 'economic voting'. As we have seen, in previous election cycles, the performance of the economy and the economic expectations of the electorate have been good guides to the popularity of the parties. Between 1992 and 1997, however, it proved difficult to account for the patterns of party popularity in these terms. No empirical version of the so-called feel-good factor and none of the traditional economic indicators appeared to correlate at all well with changes in the level of government popularity.[26] From about 1994 the standard indicators suggested that the economy was improving. The gross domestic product finally exceeded its 1990 level in 1994 and continued to grow thereafter. Inflation rates and mortgage rates were consistently low and real personal disposable income increased. Moreover, although frequent alterations to the way in which unemployment was measured gave rise to some scepticism about the reliability of the figures, there was a significant decline in the numbers out of work – from a peak of just under three million in December 1992 to 1.75 million by February 1997. Throughout 1996 the monthly figures for unemployment, GDP and real personal disposable income all continued to show steady improvement.

Governments in the past would have given a lot to achieve this sort of economic performance in the year before an election, but in this case the apparent improvement in the economy completely failed to rescue the Conservatives. This may have been because the improved economic performance was not actually felt by the voters. Thus, although unemployment fell many of the new jobs created were part-time and low level. Many of those in employment were said to feel insecure as they were increasingly employed on short-term contracts, and the threat of redundancy was ever-present. On the other hand, the evidence of the NOP/BBC exit poll suggests that a proportion of the electorate did perceive an improvement in the economy and were prepared to give the government the credit for it. More voters thought that the economy was stronger (35 per cent) rather than weaker (30 per cent) than it had been in 1992 and Labour's lead as the party most trusted to deal with the economy (45 per cent to 42 per cent) was narrower than on any other issue about which respondents were asked.

These relatively favourable perceptions did not translate into voting support, however. As Table 9 (p.214) shows, only 66 per cent of those who thought that the economy was stronger voted Conservative while those who

TABLE 9
ECONOMIC PERCEPTIONS AND VOTE IN 1997

Compared with 1992 the economy in 1997 is:

	Stronger %	Weaker %	The same %
Conservative	66	4	20
Labour	16	71	53
Liberal Democrat	13	18	20
	(815)	(707)	(779)

Party most trusted to handle the economy:

	Conservative %	Labour %	Lib Dem %
Conservative	75	2	1
Labour	12	88	10
Liberal Democrat	9	5	80
	(913)	(971)	(289)

Source: NOP/BBC exit poll.
Note: Columns do not total 100 because votes for others are not shown. Figures in brackets
indicate the number of cases on which the percentages are calculated.

thought that the economy was weaker voted overwhelmingly against the government. Similarly, a smaller proportion of those who trusted the Conservatives most on the economy actually voted for the party as compared with voters who nominated the other parties as best on the economy. Clearly, then, a significant number of electors who might have been expected to vote Conservative on economic grounds did not do so. In these cases, it must be assumed, other considerations – leadership, other policy areas, the general desirability of a change of government – overrode any sense the electorate had of 'feeling good' as a result of improved economic performance and prospects.

The election of a new government was greeted with much euphoria in left-of-centre circles and the scale of Labour's victory appeared to leave little doubt that the electorate was in the mood for change. However, a second paradoxical element of the election was that this result was achieved on the lowest turnout in any general election since 1945 (71.6 per cent). Moreover, the low turnout was not a consequence of abstentions by disillusioned Conservatives. Across constituencies, the fall in turnout was smaller the larger the previous Conservative share of the vote had been; by contrast, the larger the Labour vote in 1992, the steeper was the decline in turnout. Labour won the election with a substantial increase in its vote share, yet the turnout was lowest, and fell most sharply, in Labour areas. Constituency results suggest that there was a tendency for past and potential Labour supporters to stay away from the polls in larger numbers. This may have been because of a belief that the election was 'in the bag' or of disillusion as a consequence of Labour's abandoning traditional policy stances and ideology. Whatever the reason, it can be concluded that, although the election results can be seen as a damning indictment of the Conservative government, they hardly constituted a ringing endorsement of the new Labour Party.[27] More generally, the turnout figures for 1997 show that there is a substantial section of the electorate – concentrated in poor, inner city areas, but not confined there – whose engagement with democratic politics is slight. Whether this indicates disaffection, lack of interest or a pragmatic response to the alternatives offered is not clear, but this is certainly a problem which ought to give pause for thought to all of the parties.

CONCLUSION

It would be foolish to read too much into the results of one general election, but the electoral experience of the years since 1992 in Britain clearly supports the view that the modern electorate is a dealigned electorate. This has important consequences for politicians. The tide that flowed for Labour in 1997 could ebb just as dramatically as circumstances change. Failure to deliver on campaign pledges, policy disasters and any signs of incompetence in running the economy will be punished by an unforgiving and demanding electorate. It can no longer be assumed that any mid-term slump in popularity will eventually be reversed as old loyalties exert their influence, and there is a greater chance of the party system fragmenting as a result of a major advance by minor parties. In Scotland and Wales any disillusion with Labour is likely to be expressed in increased support for the SNP and Plaid Cymru, both of which are already securely established in terms of voting support and parliamentary representation. What might happen in England is less clear but if the Conservatives fail to improve their standing among the voters then the Liberal Democrats are in a good position to build on the 46-seat bridgehead that they established in 1997.

There are consequences too for students of electoral behaviour. Although charting the social bases of party choice remains of interest, explaining electoral behaviour in a period of rapid change demands even more emphasis than has been given hitherto on short-term influences. There remains a great deal to be done on such topics as the processes of political communication, how party leaders impinge on voters' preferences, how issue opinions are formed and develop, the impact of local and national campaigning, and the relationship between party popularity and economic performance. Given the economic contexts in which the last two elections have been fought, the conventional wisdom that 'It's the economy, stupid'[28] clearly requires re-examination. In general, it can certainly be concluded that while dealignment has made elections more unpredictable, more exciting and more paradoxical, it has also made explaining election outcomes more complicated and more difficult.

NOTES

1. A. Heath, R. Jowell and J. Curtice, with B. Taylor, *Labour's Last Chance? The 1992 Election and Beyond* (Aldershot: Dartmouth 1994) p.281.
2. R. Rose, 'Structural Change or Cyclical Fluctuation', *Parliamentary Affairs* 45/4 (Oct. 1992) pp. 451–65.
3. For a recent examination of comparative data on this question see H. Norpoth, 'The Economy', in L. LeDuc, R. Niemi and P. Norris (eds.) *Comparing Democracies: Elections and Voting in Global Perspective* (London: Sage Publications 1996) pp.299–318.
4. See B. Sarlvik and I. Crewe, *Decade of Dealignment* (Cambridge: CUP 1983).
5. The 'Alford' index is calculated by subtracting Labour's percentage share of the vote among non-manual (middle-class) workers from its share among manual (working-class) workers. Thus, if there were no difference between classes in terms of Labour support, the score would be nought; if all of the working class and none of the middle class voted Labour, the score would be 100.
6. See D. Denver, *Elections and Voting Behaviour in Britain*, 2nd ed. (Hemel Hempstead: Harvester Wheatsheaf 1994) pp.60–2.
7. See A. Heath, R. Jowell and J. Curtice, *How Britain Votes* (Oxford: Pergamon Press 1985) pp.13–16.
8. Heath *et al.* (note 1) pp.281–4.
9. See I. Crewe, 'Has the Electorate Become Thatcherite?', in R. Skidelsky (ed.) *Thatcherism* (London: Chatto & Windus 1988).
10. See D. Sanders, 'Why the Conservative Party Won – Again', in A. King (ed.) *Britain at the Polls 1992* (Chatham, NJ: Chatham House 1992) pp.206–7.
11. K. Newton, 'Economic Voting in the 1992 General Election', in D. Denver, P. Norris, D. Broughton and C. Rallings (eds.) *British Elections and Parties Yearbook 1993* (Hemel Hempstead: Harvester Wheatsheaf 1993).
12. See, e.g., H. Norpoth, *Confidence Regained: Economics, Mrs Thatcher and the British Voter* (Ann Arbor: U. of Michigan Press 1992).
13. D. Sanders, H. Ward, and D. Marsh, 'Government Popularity and the Falklands War: a Reassessment', *British Journal of Political Science* 17 (1987) pp.281–313. Sanders, 'Why the Conservatives Won – Again', in King (note 10).
14. D. Sanders, 'Government Popularity and the Next General Election', *Political Quarterly* 62 (1991) pp.235–61.
15. I. Crewe, 'Why the Conservatives Won'; in H. Penniman (ed.) *Britain at the Polls 1979* (Washington DC: American Enterprise Institute 1981) p.275.
16. See A. Mughan, 'Party leaders and Presidentialism in the 1992 Election: a Post-War Perspective', in Denver *et al.* (note 11).
17. See Crewe (note 15).
18. W. L. Miller, *Media and Voters* (Oxford: Clarendon Press 1991) Ch.8.
19. J. Curtice and H. Semetko, 'Does it Matter what the Papers Say?', in Heath *et al.* (note 1).

20. The Pedersen index is calculated by summing the changes in each party's share of the vote between elections (ignoring signs) and dividing the result by two.

21. P. Norris, *Electoral Change Since 1945* (Oxford: Blackwell 1997) p.114.

22.. For further discussion of the effect of 'Black Wednesday' on public opinion see I. Crewe, 'Electoral Behaviour', in D. Kavanagh and A. Seldon (eds.) *The Major Effect* (London: Macmillan 1994) pp.107–10.

23. The causes of the government's electoral unpopularity are discussed at greater length in D. Denver, 'The Government that Could Do No Right' in King (note 10).

24. I am grateful to Nick Moon of NOP and Brian Horrocks of the BBC for the early release of the exit poll and for allowing academic commentators to use the results.

25. It is well established by panel studies that when voters are asked to recall their party choice in a previous election they have a tendency to report that it was consistent with their current preference, even when it was not.

26. For a particularly accessible account of the first half of the 1992–97 electoral cycle in these terms see D. Sanders, '"It's the Economy Stupid": The Economy and Support for the Conservative Party, 1979–94', *Talking Politics* 3 (1995) pp.158–67.

27. A full analysis of turnout in the 1997 election is given in D. Denver and G. Hands, 'Turnout', *Parliamentary Affairs* 50 (Oct. 1997) pp.720–32.

28. Expression taken from Sanders (note 26).

Abstracts

Britain in the Nineties: The Politics of Paradox
An Introduction
HUGH BERRINGTON

This introduction discusses the various themes raised by the authors of the eight essays in this volume and, in particular, emphasises the widespread element of paradox in contemporary British politics. It looks at changes in the political parties, the impact of Europe as an issue and the elusive nature of Thatcherism. It considers the new activism of the courts and the new regulatory institutions in the privatised utilities. In addition, it reflects on the contradictions displayed by recent developments in the periphery and in the 1997 general election. It concludes with a short appraisal of electoral reform.

Power in the Parties: R. T. McKenzie and After
DENNIS KAVANAGH

The distribution of power in British political parties has been the subject of intense debate amongst practitioners and students of politics. R. T. McKenzie claimed that, contrary to Labour's self-portrait of internal democracy, its party leaders were as dominant as their Conservative counterparts. In recent years, however, it is the Conservative Party which has proved unruly and Labour which has become more leader-dominated. This contribution explains and analyses how the apparent paradox has come about.

Europe, Thatcherism and Traditionalism: Opinion, Rebellion and the Maastrict Treaty in the Backbench Conservative Party, 1992–1994
HUGH BERRINGTON and ROD HAGUE

This analysis investigates the rebellion over the Maastricht Treaty within the parliamentary Conservative Party, using a range of evidence including Early Day Motions. Past studies have tended to emphasise the Thatcherite free-market character of the revolt; we stress, however, the traditionalist roots of opposition to Maatstricht, noting a strong association between support for capital and corporal punishment and hostility to the Treaty. The article examines the relationship between opinion and division-lobby behaviour, the link between demographic variables (age, educational background, constituency characteristics, etc.) and the pattern of attitudes towards Europe within the backbench Conservative Party.

From Hostility to 'Constructive Engagement': the Europeanisation of the Labour Party
PHILIP DANIELS

The Labour Party's repositioning on the European issue has been one of the most important recent changes in British party politics. The party has moved incrementally from a policy of hostility to Britain's membership of the EC in the early 1980s to one of positive support for European integration by the end of that decade. In the 1990s, the Labour Party has effectively supplanted the Conservatives as the leading pro-European party. Labour's 'Europeanisation' has been shaped by the interplay of electoral calculations, changes in party and trade union thinking, and the dynamics of the European integration process.

Narratives of 'Thatcherism'
MARK BEVIR and R.A.W. RHODES

This essay explores the legacy of 'Thatcherism' and how the notion has been constructed by the dominant traditions of British government. We introduce the Tory, Liberal, Whig and Socialist traditions, arguing that each produces distinctive narratives – or maps, questions, languages and historical stories – about 'Thatcherism'. We discuss these several narratives to show there is no unified, monolithic, essentialist account. There is no such thing as 'Thatcherism' because each tradition constructs its own version. All we can say is that 'Thatcherism' highlighted the political salience of certain dilemmas for all of these traditions. The key dilemmas focus on welfare dependence, overload, inflation and globalisation. These dilemmas forced a reconsideration of the existing beliefs of each tradition. 'Thatcherism' lives on, therefore, not as a single narrative, but in the changes it has brought to the evolving traditions of British government.

Institutions, Regulation and Change: New Regulatory Agencies in the British Privatised Utilities
MARK THATCHER

After privatisation, new regulatory agencies were established in the British utilities. An institutional analysis is used to examine two sets of paradoxes that have arisen. First, why and how the agencies have achieved their central powerful position, despite being established by Conservative governments pledged to 'deregulation' and in a country traditionally classified as a 'weak' state. Second, at the theoretical level, why the new institutional framework has facilitated change, rather than operating as a constraint on policy modification. The essay argues that the new regulatory agencies have deployed their institutional powers and resources in ways unforeseen at the time of their creation to develop wide-ranging regulation and an increase in the power of public policy makers. The new institutional framework, in creating new, specialist regulatory agencies and in providing them with powers and flexibility, has aided and encouraged changes in the role and capacities of the state.

The Judicial Dimension in British Politics
NEVIL JOHNSON

Since the 1960s and the simplification of procedures for judicial review in the early 1980s, judicial interventions have increased and the courts have developed further the principles they apply to control executive action. Judges have also been increasingly employed in an investigative role and recently this has brought some of them into the crossfire of political controversy. The process of judicialisation has been influenced too, by Britain's increasing dependence on European jurisdictions. Overall a shift from a reliance on political discretion to an emphasis on formal rules and individual rights has occurred. This probably owes more to changes in social attitudes and public expectations than to the conduct of governments in the period since 1979.

The Periphery and its Paradoxes
WILLIAM L. MILLER

Paradoxes are manifest absurdities which none the less contain at least a grain of truth. Focusing especially, but not exclusively, on Scotland, this essay reviews more than a dozen manifestly absurd propositions about territorial politics in Britain, searching for the grains of truth that would elevate them to the status of paradoxes. They range widely across matters that concern the constitution, all of the major parties, and even the public themselves: the 'West Lothian question', tax-raising powers and the electoral system; sovereignty, subsidiarity and separatism; unionism, nationalism, and devolution; national culture and identity. Their status as paradoxes, rather than mere absurdities, depends upon the perspective of the reader as well as on the contradictions within politics and politicians themselves, so it is unlikely that there will be a complete consensus on which are paradoxes and which are mere absurdities, but it is unlikely that any reader will dismiss them all as mere absurdities.

The British Electorate in the 1990s

DAVID DENVER

Given that the modern British electorate is generally characterised as volatile and unpredictable, Conservative domination of elections from 1979 to 1992 appears paradoxical. It partly resulted from long-term changes in the social composition of the electorate but can also be explained, in ways consistent with dealignment theory, by reference to the Conservatives' handling of the economy, the popularity of their leaders and the support that they received from the tabloid press.

On the other hand, the Labour landslide in 1997 seems to vindicate the thesis of dealignment. But this result gives rise to further paradoxes. Most notably, the widely-held view, that governing parties win elections if economic performance and prospects are good, was undermined by the results of the 1997 election.

About the Contributors

Hugh Berrington is Emeritus Professor of Politics at the University of Newcastle-upon-Tyne and former President of the Political Studies Association. He was Editor of *Change in British Politics* (Frank Cass 1984) and author or co-author of other books, especially on backbench opinion. He has a special interest in the psychology of politicians.

Mark Bevir is Lecturer in Politics at the University of Newcastle-upon-Tyne and author of *The Logic of the History of Ideas* (Cambridge University Press, 1998).

Philip Daniels is lecturer in European Politics, Department of Politics, University of Newcastle-upon-Tyne, co-author of *Comparative Government and Politics*, and contributed to *Change in British Politics.*

David Denver is Professor of Politics at Lancaster University. His books include *Modern Constituency Electioneering* (1997) and *Elections and Voting Behaviour in Britain* (1994).

Rod Hague is a Senior Lecturer in Politics at the University of Newcastle-upon-Tyne and co-author of *Comparative Government and Politics* (Macmillan 1992) and contributed to *Change in British Politics* (1984).

Nevil Johnson, is Emeritus Fellow at Nuffield College, Oxford. He was Reader in the Comparative Study of Institutions and has published on British government and constitutional matters.

Dennis Kavanagh is Professor of Politics at Liverpool University. His recent books include *The British General Election of 1997* (Macmillan 1997), *The Broadening of British Politics After Thatcher* (OUP 1997) and *Election Campaigning: The New Marketing of Politics* (Blackwell 1996).

William L. Miller is Edward Caird Professor of Politics at the University of Glasgow, and a Fellow of the British Academy. His two most recent books are *Political Culture in Contemporary Britain* (with Annis May Timpson and Michael Lessnoff: OUP 1996) and *Values and Political Change in Postcommunist Europe* (with Stephen White and Paul

Heywood: Macmillan Press 1998). With Tatyana Koshechkina and Ase Grodeland he is currently researching how officials behave towards citizens in postcommunist Europe.

R.A.W. Rhodes is a Professor of Politics at the University of Newcastle-upon-Tyne; Director of the ESRC Whitehall Programme; and editor of Public Administration. His most recent books include *Understanding Governance* (Open University Press, 1997); and (with P. Weller and H. Bakviss, eds.) *The Hollow Crown* (Macmillan 1997).

Mark Thatcher, Department of Government, London School of Economics and Political Science.

Index

Books of Related Interest

National Parliaments and the European Union

Philip Norton, *University of Hull* (Ed)

'...to anyone seriously interested in the question of which Parliament should be legislating for Europe the book is a must'

The European

National parliaments are central to the political systems of Western Europe. Yet what role have they played in the development of the European Union? This volume is the first to study how national parliaments have adapted to the effects of the Single European Act and the Maastricht Treaty. Detailed studies of ten national parliaments lead to conclusions as to how they have adapted, or not adapted to European integration.

204 pages 1996 0 7146 4691 1 cloth 0 7146 4330 0 paper
The Library of Legislative Studies
A special issue of The Journal of Legislative Studies

Austen Chamberlain and the Commitment to Europe

British Foreign Policy 1924–29

Richard S Grayson, *Open University*

'Grayson should be congratulated on his clarity of purpose and lucidity.'

The Reformer

This book fills a major gap in the study of inter-war British foreign policy: It is the first complete study of Austen Chamberlain's term of office as Stanley Baldwin's Foreign Secretary from 1924–29. Overall, Chamberlain is shown to have committed Britain to a European diplomatic role, which was opposed by Cabinet ministers who did not see a European interest to all aspects of British foreign policy. Today in the Conservative Party the debate is still unresolved.

336 pages 1997 0 7146 4758 6 cloth 0 7146 4319 X paper

Britain's Failure to Enter the European Community 1961–63

The Enlargement Negotiations and Crises in European, Atlantic and Commonwealth Relations

George Wilkes (Ed)

Recently-released archival material has given historians the opportunity to re-examine the impact of the initial British application to join the European Community on the development of the EC, and on the role of Britain and its European neighbours in broader world politics.

The essays collected here outline a number of factors which made the EC too young to be able to assimilate Britain's important interests, and the British over-optimistic in their approach to negotiations with the Community.

288 pages 1997 0 7146 4687 3 cloth 0 7146 4221 5 paper

Modern Constituency Electioneering
Local Campaigning in the 1992 General Election
David Denver and **Gordon Hands**, *Lancaster University*

'The book is a model of social science research and states its conclusions modestly but persuasively...the definitive study on the subject'
EPOP Newsletter

In this first major study of grass-roots election campaigning for thirty years, David Denver and Gordon Hands survey the evolution of campaigning over the past century and describe in detail how the parties organised their constituency campaigns in the 1992 election. The study – funded by the Economic & Social Research Council – examines and evaluates the campaign techniques that are now employed, and looks in detail at the role of local media and national party organisations.

368 pages 1997 0 7146 4789 6 cloth 0 7146 4345 9 paper

Regional Dynamics
The Basis of Electoral Support in Britain
William Field, *Georgian Court College, Lakewood*

'This is a pioneering book...' **Vernon Bogdanor, Oxford University**
'This comprehensive study of British and American Think Tanks tells us how they work, how they're used and hot to distinguish between them.'
Austrian Broadcasting Corporation

Many have noticed the 'North-South divide' in British politics. In this book, William Field points out that this divide marks the resurgence of a core-periphery cleavage which was also dominant in British politics in the years before 1914. He shows how astonishingly similar the geographical pattern of the vote was in the general election of 1987 to that in the two general elections of 1910, the last before the outbreak of the First World War. Many of the same constitutional issues – devolution and reform of the second chamber – were coming to the fore then as now.

224 pages 1997 0 7146 4782 9 cloth 0 7146 4336 X paper

British Elections & Parties Review Volume 7
Charles Pattie, *University of Sheffield*, **David Denver**, *University of Lancaster*, **Justin Fisher**, *London Guildhall University*, and **Steve Ludlam**, *University of Sheffield* (Eds)

British Elections & Parties Review Volume 7, continues the commitment to the publication of front-rank research on parties, elections and voting behaviour in Britain. Issues covered include key political issues for 1990s Britain: the reform of the Labour party; the use of opinion polls; the impact of the media; European integration; Scotland and regional trends; and the bases of party support.

304 pages 1997 0 7146 4860 4 cloth 0 7146 4417 X paper

Ideas and Think Tanks in Contemporary Britain

Michael David Kandiah, *Institute of Contemporary British History*, and **Anthony Seldon**, *Founding Director of the Institute of Contemporary British History* (Eds)

'It is a useful to have a book where the search for order does not dominate.'
Parliamentary Affairs

These volumes explore the influence of ideas and think tanks in contemporary Britain. Notable commentators such as Rodney Barker (Department of Government, London School of Economics and Political Science) and Andrew Gamble (Sheffield University) contemplate how ideas have shaped politics and society. The purveyors of ideas for change, the think tanks, are examined in a series of studies; and leading academics and participants' views are recorded in a number of interviews. Volume 2 includes a chapter on Demos, one of the newest and most media-friendly think tanks in Britain.

Volume 1
216 pages 1996 0 7146 4743 8 cloth 0 7146 4301 7 paper

Volume 2
216 pages 1997 0 7146 4771 3 cloth 0 7146 4328 9 paper

The Regional Dimension of the European Union

Towards a Third Level in Europe?

Charlie Jeffery, *University of Birmingham* (Ed)

The 1990s have seen intense debates about the role of regions in European integration. Changes in EU structural funding rules, the innovations of the Maastricht treaty, and the growing importance of federal and regional government within EU member states have all boosted the significance of regional tiers of government in EU politics. Taken together their effect has been to shift the balance of decision-making responsibility within the EU to a third (regional) level of government emerging in the EU policy press alongside the first (union) and second (nation-state) levels. As a result, a system of multi-level governance can increasingly be identified, in which different levels of government adopt different roles in different fields or phases of the European policy process.

224 pages 1997 0 7146 4748 9 cloth 0 7146 4306 8 paper
Cass Series in Regional and Federal Studies Volume 2
A special issue of the journal Regional and Federal Studies